T0214710

Synthesis Lectures on Engineering, Science, and Technology

The focus of this series is general topics, and applications about, and for, engineers and scientists on a wide array of applications, methods and advances. Most titles cover subjects such as professional development, education, and study skills, as well as basic introductory undergraduate material and other topics appropriate for a broader and less technical audience.

Horst Beyer

The Reasoning of Quantum Mechanics

Operator Theory and the Harmonic Oscillator

 Springer

Horst Beyer
University of Tübingen
Tübingen, Germany

Neues Gymnasium Leibniz
Stuttgart, Germany

Johannes-Kepler-Gymnasium
Stuttgart, Germany

ISSN 2690-0300 ISSN 2690-0327 (electronic)
Synthesis Lectures on Engineering, Science, and Technology
ISBN 978-3-031-17179-6 ISBN 978-3-031-17177-2 (eBook)
https://doi.org/10.1007/978-3-031-17177-2

This Springer imprint is published by the registered company Springer Nature Switzerland AG
The registered company address is: Gewerbestrasse 11, 6330 Cham, Switzerland

Contents

List of Figures

Introduction

1

It was von Neumann's insight that Heisenberg's abstract approach to quantum mechanics, which eventually lead to its standard formulation, finds its mathematical support in operator theory, in the theory of densely-defined, linear and self-adjoint operators (DLSO's) in Hilbert space. In particular, the description and interpretation of the result of the measurement process in quantum theory are closely linked to the spectral theorems for DLSO's (Fig. 1.1).

As a consequence of von Neumann's insight, in addition to its solid experimental basis, quantum theory acquired a subtle and deep mathematical foundation. Roughly speaking, it can be said that quantum theory is standing on 2 solid legs, the mathematics of DLSO's and the experiment.

Regrettably, von Neumann's insights are hardly visible in standard quantum theory textbooks, thereby obstructing a complete understanding of the theory. For example, unlike indicated in standard textbooks, an observable in quantum theory needs to be a DSLO, not just a DLO, i.e., a densely-defined, linear and "Hermitian" operator.

Roughly speaking a DLO is an operator that can be moved inside the scalar product of a real or complex Hilbert space, from right to left without change. A DSLO is a DLO whose adjoint coincides with the DLO. Self-adjointness is a strong property that guarantees that the spectrum of the operator in question is part of the real numbers and the availability of spectral theorems that are crucial for the interpretation in quantum theory. The existence of non-real spectral values is not acceptable for observables.[1] In addition, with every formal partial differential operator, i.e., an operator that maps sufficiently differentiable functions into other functions, there comes the question of which function space to use for its representation. The choice of representation space crucially affects the properties of the resulting operator. In quantum field theory, it is very well known that there are in-equivalent representations of the algebra of canonical commutation rules, leading to physically different theories.[2] So,

---

[1] See Sect. A.3.4.

[2] See, e.g., [1–8].

H. Beyer, *The Reasoning of Quantum Mechanics*, Synthesis Lectures on Engineering, Science, and Technology, https://doi.org/10.1007/978-3-031-17177-2_1

Fig. 1.1 Quantum theory is
standing on 2 solid legs, the
mathematics of DSLO's and
the experiment

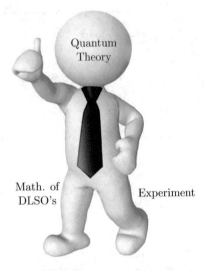

at least in the case of physical systems with infinitely many degrees of freedom, like those
considered in field theory, even this representation step is not trivial.

Another example of a common misconception is indicated by the discussion around
the "the collapse of the wave function," due to the process of a measurement. That wave
functions are not observable is an essential part of quantum mechanics. A collapse of an
object that is not observable appears to be of no relevance to physics. So, at the very least,
the appearance of the term "wave function" in the name of this discussion is misleading.
In addition, even in classical mechanics, the process of measurement results in a qualitative
change in the information on the physical system, due to excluding a continuum of possible
values through the measurement of a physical quantity. This is not essentially different in
quantum mechanics. In any case, in connection with all these discussions it is important to
know in detail what the theory is actually claiming, and also in this the mathematical basis
is relevant. In addition, there was no apriori reason why the largely separate development
of quantum mechanics and operator theory had to lead to a mathematical foundation for
quantum mechanics. It is the opinion of the author that this fact adds credit to the theory, in
addition to the overwhelming experimental evidence.

The main intention of the present volume is to merge the theory of DSLO's into the process
of the quantization of a physically relevant mechanical system, the harmonic oscillator, in this
way giving an example of how the process of quantization and subsequent treatment of the
quantum system can be carried out in any quantum mechanical textbook, in a mathematically
rigorous way and without losing the physics.

We alert readers from the field of mathematics that, contrary to the standard convention in mathematics, we use in this book the standard convention in physics that scalar products are anti-linear in the first argument and linear in the second argument.

H. R. B.

Acknowledgements I am indebted to the publisher Springer Nature, in particular, to the Executive Editor Synthesis, Susanne Filler, for the great support. In addition, I am indebted to numerous colleagues, both, from the field of mathematics and physics. The graphics in this text have been created with Wolfram Mathematica® software (www.wolfram.com) and PGF/TikZ software. The text was produced, using the document preparation system LaTeX.

H. R. B.

References

1. Baez J C, Segal I E, Zhou Z 1992, *Introduction to algebraic and constructive quantum field theory*, Princeton University Press: Princeton.
2. Bjorken J D, Drell S D 1965, *Relativistic quantum fields*, McGraw-Hill: New York.
3. Buchholz D 2000, *Algebraic quantum field theory: A status report*, Plenary talk given at XII-Ith International Congress on Mathematical Physics, London, http://xxx.lanl.gov/abs/math-ph/0011044.
4. Fulling S A 1989, *Aspects of quantum field theory in curved spacetime*, Cambridge University Press, Cambridge.
5. Haag R 1996, *Local quantum physics: Fields, particles, algebras*, Springer: New York.
6. Schroer B 2001, *Lectures on algebraic quantum field theory and operator algebras*, http://xxx.lanl.gov/abs/math-ph/0102018.
7. Streater R F, Wightman A S 2000, *PCT, Spin and statistics, and all that*, Princeton University Press: Princeton.
8. Wald R M 1994, *Quantum field theory in curved spacetime and black hole thermodynamics*, University of Chicago Press: Chicago.

The Classical Mechanical System

2

We consider the motion of a mass $m > 0$ on a spring, with spring constant $k > 0$. First, the spring stretches to balance the gravity. The corresponding position of the mass is referred to as equilibrium position. The vertical displacement of the mass from its equilibrium position is indicated by $q \in \mathbb{R}$, i.e, a positive value of q corresponds to a stretched spring, whereas a negative value of q corresponds to a compressed spring. The equilibrium position is indicated by $q = 0$. In the following, we refer to this system briefly as *"harmonic oscillator."*

Newton's equation of motion for the harmonic oscillator is given by[1]

$$mq''(t) = -kq(t) \,, \qquad (2.1)$$

for every $t \in \mathbb{R}$, resulting in

$$q(t) = q_0 \cos(\omega(t - t_0)) \,,$$

all $t \in \mathbb{R}$, where $'$ denotes the time derivative,

$$\omega := \sqrt{\frac{k}{m}} \,,$$

and $q_0 \in \mathbb{R}$ is the position of the mass at time $t_0 \in \mathbb{R}$ (Figs. 2.1 and 2.2).

---

[1] The reader might suspect that force of gravity or "weight" of the mass is missing on the right hand side of (2.1), but this is taken into account by the choice of the variable q. Another form of Newton's equation of motion for the system of the mass and the spring, under the influence of gravity, is given by $mx''(t) = -kx(t) + mg$, where x denotes the elongation of the spring from its unstressed length and g denotes the gravitational acceleration, of about 9.81 m/s^2. The system is in equilibrium, for the elongation x_0 of the spring satisfying $kx_0 = mg$. Hence, by defining the new variable $q := x - x_0$, we arrive at (2.1), which is again of the form of Newton's equations of motion.

© The Author(s), under exclusive license to Springer Nature Switzerland AG 2022
H. Beyer, *The Reasoning of Quantum Mechanics*, Synthesis Lectures on Engineering, Science, and Technology, https://doi.org/10.1007/978-3-031-17177-2_2

Fig. 2.1 Sketch of mass $m > 0$ on a massless spring, with spring constant $k > 0$, a simple example of a harmonic oscillator. The dotted circle indicates the position for the mass corresponding to a stretched spring

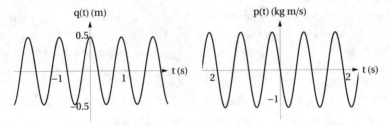

Fig. 2.2 Graph of the position q and momentum p as a function of time, for a mass $m = 0.4$ kg on a spring, with spring constant $k = 18.33$ N/m. Here, $p(0) = p_0 = 0$, $q(0) = q_0 = 0.475$ m, $t_0 = 0$

We note that

$$\left(\frac{m}{2}q'^2\right)' = m\,q'\cdot q'' = -k\,q\cdot q' = -\left(\frac{k}{2}q^2\right)' ,$$

implying that

$$\left(\frac{m}{2}q'^2 + \frac{k}{2}q^2\right)' = 0 .$$

Hence the corresponding conserved energy E is given by

$$E = \frac{m}{2}q'^2 + \frac{k}{2}q^2 = \frac{p^2}{2m} + \frac{kq^2}{2} , \qquad (2.2)$$

where p is the momentum function corresponding to q, given by

$$p = mq' .$$

Note that the energy E can assume every value of the closed interval from 0 to ∞, $[0, \infty)$. *Inside the subsequent Section, we are going to see that the corresponding quantum system can assume only a discrete set of positive energies, instead.* On the other hand, the momentum and the position of the quantum system can assume all real numbers, as is the case for the classical system.

The phase space of the system, i.e., the space of all possible values of momenta p and positions q, is given by $\mathbb{R}^2$. *Also, p and q are conjugate variables.* From (2.2), we can read off that the Hamiltonian function H of the system is given by

$$H(p, q) := \frac{p^2}{2m} + \frac{k}{2}q^2 = \frac{p^2}{2m} + \frac{m\omega^2}{2}q^2 \,,$$

for every $(p, q) \in \mathbb{R}^2$.

The Corresponding Quantum System

3

In the following, we are going to "quantize" the classical system, i.e., provide a corresponding quantum version of the classical system. For this purpose, 2 things are needed, a state space, comprising the possible physical states of the quantum system, and a representation of the observables of the classical system as DSLO's, satisfying the canonical commutation rules. Before we proceed to the quantization of the classical system, we insert 2 subsections that prepare the ground for the quantization.

3.1 Insert 1: Spectral Representation and Functional Calculus for Self-Adjoint Operators on $\mathbb{C}^n$, Equipped with the Canonical Scalar Product, $\langle\,|\,\rangle_c$

The purpose of this subsection is to become acquainted with spectral representations of operators and corresponding functional calculi, in the relatively simple case of linear and self-adjoint operators on $\mathbb{C}^n$. Such operators can be identified with Hermitian matrices. On the other hand, for this representation an auxiliary mathematical object, a basis of $\mathbb{C}^n$ needs to be selected, which is arbitrary. Although infinite dimensional matrices initially played an important role in the creation of quantum mechanics, in the description of observables, in the abstract approach of Heisenberg, Born and Jordan, see [1, 2], an approach that is referred to as "matrix mechanics," the use of such matrices is nowadays uncommon in quantum theory. Nowadays, the primary object of study is the linear map itself, which is equally abstract. This approach is also pursued in the following. On the other hand, it should not come as surprise that, since differently to classical mechanics that describes objects that are accessible to our perception, quantum mechanics describes objects that are inaccessible to our perception, without special instruments, requires also significantly more abstract mathematics. The range of atomic distances, which is of the order of 1 Å, i.e., 10^{-8} cm, is inaccessible to human perception.

© The Author(s), under exclusive license to Springer Nature Switzerland AG 2022
H. Beyer, *The Reasoning of Quantum Mechanics*, Synthesis Lectures on Engineering,
Science, and Technology, https://doi.org/10.1007/978-3-031-17177-2_3

Essentially, the following is a reformulation, of what the reader might have learned in a Linear Algebra or Analytic Geometry course in college, with a view towards quantum theory.

For densely-defined, linear and self-adjoint operators in Hilbert spaces, quite similar spectral representations and functional calculi exist, in this, integration with respect "Lebesgue-Stieltjes measures" plays a fundamental role, in the interpretation of "infinite sums" needed in the formulation of these theorems. These replace finite sums present in the formulation of spectral representations and functional calculi for linear, self-adjoint operators on $\mathbb{C}^n$.

Proofs are given in specialized functional analysis courses.[1] *These theorems play a fundamental role in the interpretation of quantum theory.* The functional calculus corresponding to such operators will be used in some form in the following.

In the following, let $n \in \mathbb{N}^*$, $\langle | \rangle_c : \mathbb{C}^n \times \mathbb{C}^n \to \mathbb{C}$ be the canonical scalar product for $\mathbb{C}^n$, defined by

$$\langle v | w \rangle_c := v_1^* w_1 + \dots v_n^* w_n,$$

for all vectors $v = {}^t(v_1, \dots, v_n)$, $w = {}^t(w_1, \dots, w_n) \in \mathbb{C}^n$, where $v_1, \dots, v_n \in \mathbb{C}$ and $w_1, \dots, w_n \in \mathbb{C}$ are the components of the vector v and w, respectively, and $*$ denotes complex conjugation.

Further, let $A : \mathbb{C}^n \to \mathbb{C}^n$ be linear map, ("linear operator"), that is in addition self-adjoint, i.e., A is such that

$$A(v + w) = A(v) + A(w), \quad A(\lambda v) = \lambda A(v),$$

for all vectors $v, w \in \mathbb{C}^n$ and $\lambda \in \mathbb{C}$ and such that

$$\langle v | A(w) \rangle_c = \langle A(v) | w \rangle_c, \tag{3.1}$$

for all $v, w \in \mathbb{C}^n$.

The reader might remember, from a course in Linear Algebra or Analytic Geometry, that the canonical basis $e_1, \dots, e_n \in \mathbb{C}^n$ of $\mathbb{C}^n$, where for every $j \in \{1, \dots, n\}$, the components of the vector e_j are all zero, apart from the jth component that is given by 1, is orthonormal with respect to the canonical scalar product, i.e.,

$$\langle e_j | e_k \rangle_c = 0,$$

for all $j, k \in \{1, \dots, n\}$ such that $j \neq k$ and

$$\langle e_j | e_k \rangle_c = 1,$$

for all $j, k \in \{1, \dots, n\}$ such that $j = k$. As a consequence,

---

[1] For e.g., see [3–5].

$$A(v) = A(v_1 e_1 + \ldots v_n e_n) = v_1 A(e_1) + \cdots + v_n A(e_n)$$

$$= v_1 \sum_{k=1}^{n} \langle e_k | A(e_1) \rangle_c \, e_k + \cdots + v_n \sum_{k=1}^{n} \langle e_k | A(e_n) \rangle_c \, e_k$$

$$= \sum_{k=1}^{n} \left[\langle e_k | A(e_1) \rangle_c \, v_1 + \cdots + \langle e_k | A(e_n) \rangle_c \, v_n \right] e_k$$

$$= \sum_{k=1}^{n} \left[\sum_{l=1}^{n} \langle e_k | A(e_l) \rangle_c \, v_l \right] e_k = \sum_{k=1}^{n} \left(\sum_{l=1}^{n} M_{kl} v_l \right) e_k,$$

for every $v = {}^t(v_1, \ldots, v_n)$, where M_{kl} is kl-th component of the $n \times n$-matrix

$$M := \begin{pmatrix} \langle e_1 | A(e_1) \rangle_c & \cdots & \langle e_1 | A(e_n) \rangle_c \\ . & \cdots & . \\ . & \cdots & . \\ . & \cdots & . \\ \langle e_n | A(e_1) \rangle_c & \cdots & \langle e_n | A(e_n)_c \rangle \end{pmatrix}.$$

The matrix M is called the representation matrix of A with respect to the canonical basis. In particular, since A is self-adjoint, M is Hermitian, i.e.,

$$M_{lk} = M_{kl}^*,$$

for all $k, l \in \{1, \ldots, n\}$.

   In the following, wherever possible, we use the standard convention for linear maps that skips the evaluation brackets ( ), when a linear map is applied to a vector from its domain. For instance, for every $v \in \mathbb{C}^n$, we write "Av" instead of "$A(v)$." In addition, usually and as is common, we are not going to indicate the use of identical maps as far as possible. For instance, in our case in question, $A - \lambda$, where $\lambda \in \mathbb{C}$, means "A minus λ times the identical map on $\mathbb{C}^n$". See, e.g., (3.2).

   Also, we alert the reader to the fact that, since the *state spaces* in quantum theory, differently to $\mathbb{C}^n$, which is a finite dimensional vector space, are *practically always infinite dimensional*, in the description of operators from quantum theory, the use of matrices is unsuitable (clumsy) and therefore uncommon. Hence, also in the following, we will work with the operators itself, not with representation matrices. As a consequence, the following derivation of the spectral representation of A and the functional calculus for A, will give a the reader a preview of the treatment of operators in quantum theory.

   Further, let $\lambda_1, \ldots, \lambda_r$, where $r \in \{1, \ldots, n\}$, be the *pairwise different spectral values*[2] of A,

---

[2] Since $\mathbb{C}^n$ is a *finite* dimensional vector space, the spectral values of A consist of eigenvalues, namely the eigenvalues of M. Also, since M is Hermitian, these eigenvalues are all real and eigenvectors, corresponding to different eigenvalues, are orthogonal.

$$\sigma(A) := \{\lambda_1, \ldots, \lambda_r\},$$

the so called spectrum of A, and for every $i \in \{1, \ldots, r\}$, $u_{i1}, \ldots, u_{is_i}$ be an orthonormal basis of the eigenspace of A to the eigenvalue λ_i:

$$\ker(A - \lambda_i) = \{v \in \mathbb{C}^n : Av = \lambda_i v\} \ . \tag{3.2}$$

Then, *according to Linear Algebra*,

$$u_{11}, \ldots, u_{1s_1}, \ldots, u_{r1}, \ldots, u_{rs_r}$$

is an orthonormal basis of $(\mathbb{C}^n, \langle\,|\,\rangle_c)$.

Using this basis, we can derive a so called *spectral representation of A*. For this purpose, we define for every $i \in \{1, \ldots, r\}$, $P_i : \mathbb{C}^n \to \mathbb{C}^n$ by

$$P_i u = P_i \sum_{j=1}^{r} \sum_{s=1}^{s_j} \langle u_{js} | u \rangle_c \, . u_{js} := \sum_{s=1}^{s_i} \langle u_{is} | u \rangle_c \, . u_{is} \in \ker(A - \lambda_i . \mathrm{id}_{\mathbb{C}^n}),$$

for every $u \in \mathbb{C}^n$.[3] We note for every $i \in \{1, \ldots, r\}$ that P_i is linear,

$$P_i^2 := P_i \circ P_i = P_i,$$

and that

$$\langle u | P_i v \rangle_c = \langle u | \sum_{s=1}^{s_i} \langle u_{is} | v \rangle_c \, . u_{is} \rangle_c = \sum_{s=1}^{s_i} \langle u_{is} | v \rangle_c \, \langle u | u_{is} \rangle_c$$

$$= \sum_{s=1}^{s_i} \langle u_{is} | u \rangle_c^* \, \langle u_{is} | v \rangle_c = \langle \sum_{s=1}^{s_i} \langle u_{is} | u \rangle_c \, . u_{is} \big| v \rangle_c = \langle P_i u | v \rangle_c$$

for all $u, v \in \mathbb{C}^n$, i.e., that P_i is self-adjoint. As a consequence, P_i is an *orthogonal projection onto the subspace*

$$\ker(A - \lambda_i . \mathrm{id}_{\mathbb{C}^n})$$

of $\mathbb{C}^n$.

Further, we conclude that

$$Au = A \sum_{i=1}^{r} \sum_{s=1}^{s_i} \langle u_{is} | u \rangle_c \, . u_{is} = \sum_{i=1}^{r} \sum_{s=1}^{s_i} \langle u_{is} | u \rangle_c \, . A u_{is} = \sum_{i=1}^{r} \sum_{s=1}^{s_i} \langle u_{is} | u \rangle_c \, . \lambda_i . u_{is}$$

$$= \sum_{i=1}^{r} \lambda_i \sum_{s=1}^{s_i} \langle u_{is} | u \rangle_c \, . u_{is} = \sum_{i=1}^{r} \lambda_i \, P_i u = \left(\sum_{i=1}^{r} \lambda_i \, P_i \right) u$$

---

[3] For every set M, the symbol id_M denotes the map that associates with every member a of M that same member a of M.

and hence that

$$A = \sum_{i=1}^{r} \lambda_i . P_i .$$

The latter representation is called a *spectral representation of A*. It is unique up to the order of summation.

Using the spectral representation of A, we define a so called *functional calculus* for the operator A by

$$f(A) := \sum_{i=1}^{r} f(\lambda_i) . P_i,$$

for every complex-valued function f *with domain containing the spectrum of A*.

Remark 3.1.1 Note that for complex-valued functions f, g with domains containing the spectrum of A and such that for every $i \in \{1, \ldots, r\}$

$$f(\lambda_i) = g(\lambda_i),$$

it follows that

$$f(A) = \sum_{i=1}^{r} f(\lambda_i) . P_i = \sum_{i=1}^{r} g(\lambda_i) . P_i = g(A) .$$

In addition, note that

$$P_i P_j = \delta_{ij} . P_j$$

for all $i, j \in \{1, \ldots, r\}$, where $\delta_{ij} := 1$ if $i = j$ and $\delta_{ij} := 0$ if $i \neq j$, for all $i, j \in \{1, \ldots, r\}$.

Exercise 1
(a) Show that

$$1_{\mathbb{R}}(A) = \mathrm{id}_X, \ \mathrm{id}_{\mathbb{R}}(A) = A,$$

where $1_{\mathbb{R}} : \mathbb{R} \to \mathbb{R}$ is defined by $1_{\mathbb{R}}(x) := 1$ for every $x \in \mathbb{R}$.
(b) Let $i \in \{1, \ldots, r\}$ and f_i some function whose domain contains $\sigma(A)$ and is such that

$$f_i(\lambda_j) = \delta_{ij}$$

for all $j \in \{1, \ldots, r\}$. Show that

$$f_i(A) = P_i \ .$$

Solution 1 (a) According to the definitions, for every $u \in \mathbb{C}^n$

$$1_{\mathbb{R}}(A)u = \sum_{i=1}^{r} 1_{\mathbb{R}}(\lambda_i).P_i u = \sum_{i=1}^{r} P_i u = \sum_{i=1}^{r} \sum_{s=1}^{s_i} \langle u_{is} | u \rangle_c .u_{is} = u = \mathrm{id}_\chi u,$$

and

$$\mathrm{id}_{\mathbb{R}}(A) = \sum_{i=1}^{r} \mathrm{id}_{\mathbb{R}}(\lambda_i).P_i = \sum_{i=1}^{r} \lambda_i.P_i = A \ .$$

(b) For every $i \in \{1, \dots, r\}$, we conclude that

$$f_i(A) = \sum_{j=1}^{r} f_i(\lambda_j).P_j = \sum_{j=1}^{r} \delta_{ij}.P_j = P_i \ .$$

Exercise 2 Show that

$$(f + g)(A) = f(A) + g(A), \quad (\alpha.f)(A) = \alpha.f(A),$$
$$(f \cdot g)(A) = f(A) \circ g(A), \quad f^*(A) = (f(A))^*,$$

for all complex-valued functions f and g with domains containing the spectrum of A, and $\alpha \in \mathbb{C}$. On the left-hand side of the last equation, the symbol * denotes complex-conjugation, and on the right side, the adjoint.

Solution 2 For f and g with domains containing the spectrum of A, $\alpha \in \mathbb{C}$ and $u \in \mathbb{C}^n$, we conclude that

$$(f + g)(A) = \sum_{i=1}^{r}(f + g)(\lambda_i).P_i = \sum_{i=1}^{r}[f(\lambda_i) + g(\lambda_i)].P_i$$

$$= \sum_{i=1}^{r}[f(\lambda_i).P_i + g(\lambda_i).P_i] = \sum_{i=1}^{r} f(\lambda_i).P_i + \sum_{i=1}^{r} g(\lambda_i).P_i$$

$$= f(A) + g(A),$$

$$(\alpha.f)(A) = \sum_{i=1}^{r}(\alpha.f)(\lambda_i).P_i = \sum_{i=1}^{r}[\alpha.f(\lambda_i)].P_i = \alpha.\sum_{i=1}^{r} f(\lambda_i).P_i$$

$$= \alpha.f(A),$$

$$[f(A) \circ g(A)]u = f(A)[g(A)u] = f(A)\left[\sum_{i=1}^{r} g(\lambda_i).P_i u\right]$$

$$= \sum_{i=1}^{r} g(\lambda_i)f(A)(P_i u) = \sum_{i=1}^{r} g(\lambda_i)\sum_{j=1}^{r} f(\lambda_j)P_j P_i u$$

$$= \sum_{i=1}^{r} g(\lambda_i)\sum_{j=1}^{r} f(\lambda_j)\delta_{ji} P_i u = \sum_{i=1}^{r} g(\lambda_i)f(\lambda_i)P_i u$$

$$= \sum_{i=1}^{r} f(\lambda_i)g(\lambda_i)P_i u = \sum_{i=1}^{r}[f(\lambda_i) \cdot g(\lambda_i)].P_i u$$

$$= \sum_{i=1}^{r}(f \cdot g)(\lambda_i).P_i u = (f \cdot g)(A)u,$$

$$f^*(A) = \sum_{i=1}^{r} f^*(\lambda_i).P_i = \sum_{i=1}^{r}(f(\lambda_i))^*.P_i = \sum_{i=1}^{r}[f(\lambda_i).P_i]^*$$

$$= \left[\sum_{i=1}^{r} f(\lambda_i).P_i\right]^* = (f(A))^* .$$

Exercise 3 Show that

$$p(A) = \sum_{k=1}^{n} a_k.A^k,$$

where $a_0, \ldots, a_n \in \mathbb{C}$, $p : \mathbb{R} \to \mathbb{R}$ is a polynomial function defined by

$$p(\lambda) := \sum_{k=1}^{n} a_k \lambda^k$$

for every $\lambda \in \mathbb{R}$, and $A^0 := \mathrm{id}_X$ and recursively $A^{k+1} := A \circ A^k$ for every $k \in \mathbb{N}$.

Solution 3 We note that

$$p = \sum_{k=1}^{n} a_k.(\mathrm{id}_\mathbb{R})^k .$$

Hence, it follows from the Exercises 1 and 2 that

$$p(A) = \sum_{k=1}^{n} a_k.[(\mathrm{id}_\mathbb{R})^k](A) = \sum_{k=1}^{n} a_k.[\mathrm{id}_\mathbb{R}(A)]^k = \sum_{k=1}^{n} a_k.A^k .$$

Exercise 4 Let $i \in \{1, \ldots, r\}$. Using Exercise 1(b), show that

$$P_i = p_i(A),$$

where the polynomial $p_i : \mathbb{R} \to \mathbb{R}$ is defined by

$$p_i(\lambda) := (1 + \lambda - \lambda_i) \cdot \prod_{j=1, j \neq i}^{r} \frac{\lambda - \lambda_j}{\lambda_i - \lambda_j}$$

for all $\lambda \in \mathbb{R}$. *Note that, according to Exercise 3, this result allows the expression of P_i in form of a polynomial in A! For the latter, all what is needed is the knowledge of the spectral values of A.*

Solution 4 Since for every $j \in \{1, \ldots, r\}$;

$$p_i(\lambda_j) = \delta_{ij},$$

it follows from Exercise 1(b) that

$$p_i(A) = P_i .$$

Exercise 5 Using the results of Exercise 4, find the orthogonal projections onto the eigenspaces and the spectral representation of the linear self-adjoint operator $A : \mathbb{C}^3 \to \mathbb{C}^3$ defined by

$$Au := (3u_1 + 2u_2 + 4u_3, 2u_1 + 2u_3, 4u_1 + 2u_2 + 3u_3)$$

for every $u = (u_1, u_2, u_3) \in \mathbb{C}^3$. Note that the representation matrix of A, with respect to the canonical basis, is given by the 3×3-matrix

$$\begin{pmatrix} 3 & 2 & 4 \\ 2 & 0 & 2 \\ 4 & 2 & 3 \end{pmatrix} .$$

Solution 5 The spectrum of A is given by $\{8, -1\}$. The orthogonal projection P_1 onto the eigenspace corresponding to 8 is given by

$$P_1 u := \frac{1}{9}.(4u_1 + 2u_2 + 4u_3, 2u_1 + u_2 + 2u_3, 4u_1 + 2u_2 + 4u_3),$$

and the orthogonal projection P_2 onto the eigenspace corresponding to -1 is given by

$$P_2 u := \frac{1}{9}.(5u_1 - 2u_2 - 4u_3, -2u_1 + 8u_2 - 2u_3, -4u_1 - 2u_2 + 5u_3),$$

for every $u = (u_1, u_2, u_3) \in \mathbb{C}^3$. The spectral representation corresponding to A is given by

$$A = 8.P_1 - P_2 .$$

Exercise 6 Express the inverse

$$A^{-1},$$

of A from Exercise 5, in form of a polynomial in A.

Solution 6 We are going to interpolate the function

$$\left(\mathbb{R}^* \to \mathbb{R}, \lambda \mapsto \frac{1}{\lambda} \right)$$

on

$$\sigma(A) = \{8, -1\},$$

by the linear function

$$f := \left(\mathbb{R} \to \mathbb{R}, \lambda \mapsto \frac{1}{8}(\lambda - 7) \right) .$$

Then

$$A^{-1} = f(A) = \frac{1}{8}.(A - 7) .$$

Exercise 7 Express

$$\exp(tA),$$

for every $t \in \mathbb{R}$ and A from Exercise 5, in form of a polynomial in A. Note that for every $u \in \mathbb{C}^3$, the corresponding function $v : \mathbb{R} \to \mathbb{C}^3$, defined for every $t \in \mathbb{R}$ by

$$v(t) := \exp(tA)u,$$

is a solution of the system of ordinary differential equations

$$v'(t) = Av(t),$$

for every $t \in \mathbb{R}$, such that $v(0) = u$.

Solution 7 Let $t \in \mathbb{R}$. We are going to interpolate the function

$$(\mathbb{R} \to \mathbb{R}, \lambda \mapsto \exp(t\lambda))$$

on

$$\sigma(A) = \{8, -1\},$$

by the linear function

$$f := \left(\mathbb{R} \to \mathbb{R}, \lambda \mapsto \exp(8t) - \frac{1}{9}[\exp(-t) - \exp(8t)](\lambda - 8) \right) .$$

Then

$$\exp(t A) = f(A) = \exp(8t) - \frac{1}{9}[\exp(-t) - \exp(8t)].(A - 8) .$$

Exercise 8 Let f be a real-valued function with domain containing the spectrum of A and assuming values in $\{0, 1\}$. Show that $f(A)$ is an orthogonal projection.

Solution 8 Since f assumes values in $\{0, 1\}$, it follows that $f \cdot f = f$, and hence that

$$f(A) = (f \cdot f)(A) = f(A) \circ f(A) .$$

Further, since f is real-valued,

$$(f(A))^* = f^*(A) = f(A),$$

i.e., $f(A)$ is self-adjoint. Hence, $f(A)$ is an orthogonal projection.

3.2   Insert 2: Operators in Hilbert Spaces

In the following, we are going to introduce some terminology of operator theory. For more detail, the reader might refer to the Appendix or to [3–10]. For this purpose, let $(X, \langle | \rangle)$ be a complex Hilbert space which is non-trivial, i.e., contains a non-zero vector.

A *linear operator in* X is a linear map, whose domain is a subspace of X.

A *linear operator whose domain* $D(A)$ *is dense in* X, i.e., such that for every $f \in X$, there is a sequence $f_1, f_2, \ldots$ of elements in the domain that converges to f, *is called a densely-defined, linear operator ("DLO") in* X. A linear operator in X, whose domain coincides with X, is also called a linear operator *on* X.

DLO's relevant for applications are predominantly discontinuous, also called unbounded, i.e., for such a DLO, there are sequences in the domain that converge to an element of its domain, but the corresponding sequence of images does not converge.

The condition of being closed is a weaker replacement of the condition of continuity that is often achievable through extension of a given operator. If $A : D(A) \to X$ is a linear operator in X, then *we call A closed if the graph of A,*

$$G(A) := \{(f, Af) : f \in D(A)\},$$

is a closed subspace of the complex Hilbert space X^2, i.e., if for every convergent sequence $f_1, f_2, \ldots$ of elements from the domain of A, which is such that the corresponding sequence of images $Af_1, Af_2, \ldots$ is also converging, it follows that

$$\lim_{\nu\to\infty} f_\nu \in D(A) \text{ and } A \lim_{\nu\to\infty} f_\nu = \lim_{\nu\to\infty} Af_\nu .$$

Adjoint operators of DLO's, defined later, *are always closed*.

Still, the exact determination of the domains of closed DLO's can only be achieved in simple cases, e.g., operators in Hilbert spaces of square integrable functions that operate through multiplication by a function. Also, domains of DLO's induced by formal[4] linear partial differential operators with bounded coefficients are usually Sobolev spaces.

For this reason, we define the notion of closability of a DLO. If $A : D(A) \to X$ is DLO, then *we call A closable, if there is a closed extension of A*. If this is the case, it is easy to show that *then there is smallest closed extension of A*, which is *called the closure of A* and is *denoted by* $\bar{A}$. The graph $G(\bar{A})$ of the closure of A can be constructed from its graph as follows. For every $(f, \bar{A}f) \in G(\bar{A})$, there is a sequence in $(f_1, Af_1), (f_2, Af_2), \ldots$ such that

$$\lim_{\nu\to\infty} f_\nu = f \text{ and } \lim_{\nu\to\infty} Af_\nu = \bar{A}f .$$

We proceed to the definition of the adjoint operator of a DLO. If $A : D(A) \to X$ is a DLO, then

$$\{(f, g) \in X^2 : \langle g|h \rangle = \langle f|Ah \rangle \text{ for all } h \in D(A)\}$$

is a closed subspace of X^2. Moreover, it is easy to show that this subspace is the graph of a uniquely determined linear operator in X, which is called the adjoint operator of A and is denoted by A^*. By construction, A^* is closed, and we have

$$\langle A^* f|h \rangle = \langle f|Ah \rangle ,$$

for every f from the domain of A^* and every $h \in D(A)$. *We call a DLO self-adjoint, if*

$$A^* = A .$$

> Observables in quantum theory need to be self-adjoint DLO's, i.e., densely-defined, linear and self-adjoint operators, DSLO's.

We call a DLO Hermitian or symmetric, if A^* is an extension of A, i.e., if

$$G(A^*) \supset G(A) .$$

We note that Hermitian DLO's have a closed extension and therefore are closable. On the other hand, Hermitian DLO's do not always have self-adjoint extensions or do not always have physically meaningful self-adjoint extensions. In these cases, they are not suitable for use for the definition of an observable in quantum theory. E.g., this is true for the candidate

---

[4] The term "formal" refers to the absence of a natural domain.

for momentum measurement of a free particle moving on the half-line, which is a closed Hermitian DLO that has no self-adjoint extensions.[5]

As said above, the exact determination of the domains of closed DLO's can only be achieved in simple cases. For this reason, we define the notion of essential self-adjointness of a DLO. *A Hermitian DLO is called essentially self-adjoint, if its closure is self-adjoint.*

For a closed DLO, $A : D(A) \rightarrow X$, we define its spectrum $\sigma(A)$ by

$$\sigma(A) := \{\lambda \in \mathbb{C} : A - \lambda \text{ is not one-to-one and onto (i.e., not bijective)}\} \ .$$

The spectrum is a closed subset of the complex plane. If A is in addition self-adjoint, then *the spectrum* is non-empty, part of the real numbers and, if associated with an observable of a quantum system, *determines the possible outcomes of measurements.* If X is finite dimensional, the spectrum consists solely of eigenvalues, but if X is infinite dimensional, which is the generic case in quantum theory, there are frequently important spectral values of observables that are not eigenvalues. In quantum theory, Hamilton operators corresponding to atoms and molecules, besides eigenvalues that correspond to bound states, contain a continuous part inside their spectra that correspond to scattering states. Also, the spectrum of operators in quantum theory that are associated with position and momentum measurement contain no eigenvalues, and this is of course physically reasonable.

The complement

$$\mathbb{C} \backslash \sigma(A)$$

of the spectrum, which is an open subset of the complex numbers, is called *resolvent set*. For every $\lambda \in \rho(A) := \mathbb{C} \backslash \sigma(A)$, $A - \lambda$ is closed as well as bijective and hence, as a consequence of the Bounded Inverse Theorem, see Theorem A.3.3 (iv),

$$R_A(\lambda) := (A - \lambda)^{-1}$$

is continuous, also called *bounded*. Further, the *resolvent* R_A is weakly complex analytic, i.e., $\langle f | R_A(\lambda) g \rangle$ is complex analytic, also called holomorphic, for all $f, g \in X$, opening the path to the use of contour integration, for the definition of a functional calculus for A.

For DSLO's there are spectral representations and functional calculi, quite similar to those for DSLO's on finite dimensional Hilbert spaces, described in the previous subsection. These play an important role in the description and the interpretation of the measurement process in quantum theory. In this, integration with respect "Lebesgue-Stieltjes measures" replace finite sums present in the formulation of spectral representations and functional calculi for linear, self-adjoint operators on $\mathbb{C}^n$. Significantly, these measures have their support inside the spectrum of the corresponding DSLO, i.e., the complement of the spectrum is a set of measure 0, with respect to integration. Comparable theorems do not exist for Hermitian DSLO's that have no self-adjoint extensions, such as the candidate for momentum measurement of a free particle moving on the half-line. In addition, the spectrum of that candidate is given

---

[5] See Sect. A.3.4.

by the closed lower half-plane[6] of the complex plane, which is another reason for making it unsuitable as an observable, since the outcome of a measurement of a physical quantity can only be a real number. The construction of these functional calculi is relatively easy for operators in Hilbert spaces of square integrable functions that operate through multiplication by a function. Therefore, in quantum theory it is frequently tried to find representation spaces for physical observables, where the observables are described by such multiplication operators. In physics, this process is called "diagonalization of an observable," since the process resembles the process of the diagonalization of a matrix.

3.3   Quantization in General

First, it is needed a complex Hilbert space, $(X, \langle \, | \, \rangle)$, called the "state space" of the system, essentially comprising the possible states of the system.

The *(pure) states*[7] of the quantum system are *rays*[8] $\mathbb{C}^*.f$, $f \in X \setminus \{0\}$, where

$$\mathbb{C}^*.f := \{\lambda.f : \lambda \in \mathbb{C}^*\},$$

in a *complex Hilbert space*[9] $(X, \langle \, | \, \rangle)$, called a *representation space* or *state space*. Such a space is unique up to Hilbert space isomorphisms, only. *Hence its elements are not observable*[10] and in this sense "abstract."

Exercise 9   According to Linear Algebra and Analysis, $\mathbb{C}$ *equipped with the canonical scalar product* $\langle \, | \, \rangle_c : \mathbb{C}^2 \to \mathbb{C}$, defined for all $u, v \in \mathbb{C}$ by

$$\langle u | v \rangle_c := u^* \cdot v,$$

is a complex Hilbert space, where * denotes complex conjugation. Give all states in this Hilbert space. How many states are there?

Solution 9   Since

$$\mathbb{C}^*.1 = \mathbb{C}^*,$$

there is only one state in this Hilbert space, given by $\mathbb{C}^*.1$.

---

[6] See Sect. A.3.4.

[7] For a brief review of the axioms of Quantum Theory see, for e.g., [11].

[8] The use of rays in this definition becomes important in the description of particles with spin.

[9] A definition of the mathematical concept of Hilbert spaces is given later on in Definition A.2.7.

[10] In particular, "wave functions," the solutions of Schrödinger equations are *not* observable.

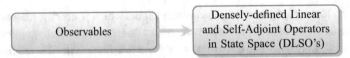

Second, we need to represent the *observables* of the classical system, i.e., momentum, position and energy, (p, q, H), *as densely-defined, linear and self-adjoint*[11] *operators*, (DLSO's), $(\hat{p}, \hat{q}, \hat{H})$, in the chosen state space. In addition, *any DLSO in the state space is considered to be an observable of the quantum system.*

Third and finally, in the latter representation, *canonically conjugate observables of the classical system* are *required to be mapped into observables satisfying canonical commutation relations (CCR)*, that are usually ill-defined, or their exponential Weylian form, that are well-defined. The latter requirement poses a strong restriction.

3.4   Concrete Quantization of a Harmonic Oscillator

3.4.1   The State Space

As representation space $(X, \langle | \rangle)$, we choose $L^2_{\mathbb{C}}(\mathbb{R})$. In the following, du denotes the Lebesgue measure in 1 dimension. We define

$$\mathcal{L}^2_{\mathbb{C}}(\mathbb{R}) := \{f : \mathbb{R} \to \mathbb{C} : \mathrm{Re}(f), \mathrm{Im}(f) \text{ are Lebesgue measurable}$$
$$\text{and } |\mathrm{Re}(f)|^2, |\mathrm{Im}(f)|^2 \text{ are Lebesgue integrable}\}$$

and for all $f, g \in \mathcal{L}^2_{\mathbb{C}}(\mathbb{R})$

$$\langle f | g \rangle := \int_{-\infty}^{\infty} f^*(u) g(u) \, du \ .$$

Then, *according to Functional Analysis,*

$(X, +, ., \langle | \rangle_2)$ is a complex Hilbert space,

where
$$X := \mathcal{L}^2_{\mathbb{C}}(\mathbb{R})/\sim,$$

the equivalence relation $\sim$ on $\mathcal{L}^2_{\mathbb{C}}(\mathbb{R})$ is defined by[12]

$$f \sim g :\Leftrightarrow f = g \text{ a.e. on } \mathbb{R},$$

---

[11] For the definition of densely-defined, linear and self-adjoint operators in Hilbert spaces, see Definition A.3.2 and Theorem A.3.4. The use of densely-defined, linear and symmetric operators, see Theorem A.3.4, for that purpose is not possible, since such operators can have non-real spectral values, disqualifying them as observables. In addition, there are no spectral theorems for such operators, which again makes them unsuitable for use as observables.

[12] A.e. stands for *almost everywhere*, i.e., $\{u \in \mathbb{R} : f(u) \neq g(u)\}$ is set of Lebesgue measure 0.

for all $f, g \in \mathcal{L}_{\mathbb{C}}^2(\mathbb{R})$, and X is *equipped with the operations* $+, .$ *and the scalar product* $\langle | \rangle$, *defined by*

$$[f] + [g] := [f + g], \quad \lambda.[f] := [\lambda.f],$$
$$\langle [f] | [g] \rangle := \langle f | g \rangle$$

for all $f, g \in \mathcal{L}_{\mathbb{C}}^2(\mathbb{R})$ and $\lambda \in \mathbb{C}$.

As is standard practice, for simplicity, in the following, we are not going to indicate anymore that we are working with equivalence classes, rather than functions. Normally, this does not lead to complications, since in applications, usually, the equivalence classes in question, have a unique distinguished, e.g., continuous, representative, which is the basis for considerations. On the other hand, occasionally, it is necessary to remember this fact.

3.4.2 Representation of the Classical Observables

For $\hat{q}$, we choose the maximal multiplication operator in X that multiplies every element of its domain $D(\hat{q})$ by the function u/k on $\mathbb{R}$, where u denotes the identical function[13] on $\mathbb{R}$ and $\kappa > 0$ is a constant with dimension 1/length that is given later.[14]

$$\hat{q} : D(\hat{q}) \to X,$$

where

$$D(\hat{q}) := \{f \in X : u \cdot f \in X\},$$

and

$$\hat{q}f := \frac{u}{\kappa} \cdot f,$$

for every $f \in D(\hat{q})$.

---

[13] So, formally, the symbol u denotes the map from the real numbers to the real numbers that associates with every real number that same number. This is a replacement of the notion of a "variable" from Calculus. In higher dimensions, "variables" from Calculus are reinterpreted as coordinate projections. Such reinterpretations of variables make classical mathematical texts legible to modern mathematicians.

[14] Note that in this case the spectrum of the position operator is given by the set of all real numbers, as is also the case in classical mechanics.

According to Operator Theory, $\hat{q}$ is a densely-defined, linear and self-adjoint operator in X, i.e., $D(\hat{q})$ is a dense subspace of X, $\hat{q}$ is a linear map and the adjoint $\hat{q}^*$ of $\hat{q}$ coincides with $\hat{q}$, i.e., if $g \in X$ is such that the linear map that associates with every $f \in D(\hat{q})$ the value

$$\langle g | \hat{q} f \rangle$$

is continuous, then $g \in D(\hat{q})$ and

$$\langle g | \hat{q} f \rangle = \langle \hat{q} g | f \rangle,$$

for every $f \in D(\hat{q})$. In addition, the spectrum $\sigma(\hat{q})$ of $\hat{q}$, i.e., the complex values λ for which $\hat{q} - \lambda$ is not one-to-one onto, i.e., for which $\hat{q} - \lambda$ is not bijective, is given by all real numbers.

$$\sigma(\hat{q}) = \mathbb{R} \ .$$

This spectrum gives the possible outcomes of measurements of the position of the particle in quantum theory. The range of theses values is the same for the classical system.
     For $\hat{p}$, we choose the closure $\frac{\hbar\kappa}{i} \overline{D_\mathbb{R}}$, of the operator $\frac{\hbar\kappa}{i} D_\mathbb{R}$, where $\hbar$ denotes Planck's reduced constant,

$$\hat{p} := \frac{\hbar\kappa}{i} \overline{D_\mathbb{R}}$$

and $D_\mathbb{R}$ is a differential operator,

$$D_\mathbb{R} : D(D_\mathbb{R}) \rightarrow X,$$

where

$$D(D_\mathbb{R}) := \{ f \in C^1(\mathbb{R}, \mathbb{C}) \cap X : f' \in X \},$$

and

$$D_{\mathbb{R}} f := f',$$

for every $f \in D(D_{\mathbb{R}})$. Here $C^1(\mathbb{R}, \mathbb{C})$ denotes the space consisting of all continuously differentiable complex-valued functions on $\mathbb{R}$.

According to Operator Theory, $\hat{p}$ is a densely-defined, linear and self-adjoint operator in X, i.e., the domain of $\hat{p}$, $D(\overline{D_{\mathbb{R}}})$, is a dense subspace of X, $\hat{p}$ is a linear map and the adjoint $\hat{p}^*$ of $\hat{p}$ coincides with $\hat{p}$, i.e., if $g \in X$ is such that the linear map that associates with every $f \in D(\overline{D_{\mathbb{R}}})$ the value

$$\langle g | \hat{p} f \rangle,$$

is continuous, then $g \in D(\overline{D_{\mathbb{R}}})$ and

$$\langle g | \hat{q} f \rangle = \langle \hat{q} g | f \rangle,$$

for every $f \in D(\overline{D_{\mathbb{R}}})$. In addition, the spectrum $\sigma(\hat{p})$ of $\hat{p}$, i.e., the complex values λ for which $\hat{p} - \lambda$ is not one-to-one onto, i.e., for which $\hat{p} - \lambda$ is not bijective, is given by all real numbers.

$$\sigma(\hat{p}) = \mathbb{R}.$$

This spectrum gives the possible outcomes of measurements of the momentum of the particle in quantum theory. The range of these values is the same for the classical system.

A candidate for the Hamilton operator $\hat{H}$, is given by[15]

$$\frac{1}{2m} \hat{p}^2 + \frac{k}{2} \hat{q}^2.$$

On the other hand, this operator is not self-adjoint, and instead of studying the self-adjoint extensions of this operator, it is advantageous to use the restriction $\hat{H}_0$ to Schwartz space, i.e., the subspace D_0 of X, defined by

---

[15] In the following, we always assume composition of maps (which includes addition, multiplication etc.) to be maximally defined. For instance, the addition of 2 maps is defined on the (possibly empty) intersection of their domains.

$$D_0 := \{ f \in C^\infty(\mathbb{R}, \mathbb{C}) \cap X : f^{(n)} \in X \wedge u^n f \in X, \text{ for every } n \in \mathbb{N}^* \},$$

as a basis for a direct definition of the Hamilton operator $\hat{H}$.

The space D_0 is dense in X, since it contains $C_0^\infty(\mathbb{R}, \mathbb{C})$, the space of infinitely often continuously differentiable complex-valued functions on $\mathbb{R}$ that vanish outside of a bounded interval. We note that

$$\hat{H}_0 f = \frac{\hbar^2 \kappa^2}{2m} \left[-f'' + \left(\frac{m\omega}{\hbar \kappa^2} \right)^2 u^2 f \right] = \hbar\omega \left(-f'' + \frac{1}{4} u^2 f \right),$$

for every $f \in D_0$, where we define

$$\kappa := \left(\frac{2m\omega}{\hbar} \right)^{1/2}.$$

We note that κ has the dimension of $1/\text{length}$.

For the definition of $\hat{H}$, we define linear operators $b : D_0 \to X$ and $b^\dagger : D_0 \to X$ in X, by

$$bf := \frac{1}{i} f' - \frac{i}{2} uf = \frac{1}{i} \exp\left(-\frac{u^2}{4} \right) \left[\exp\left(\frac{u^2}{4} \right) f \right]',$$

$$b^\dagger f := \frac{1}{i} f' + \frac{i}{2} uf = \frac{1}{i} \exp\left(\frac{u^2}{4} \right) \left[\exp\left(-\frac{u^2}{4} \right) f \right]',$$

for every $f \in D_0$, where
     Then,

$$\langle f | bg \rangle = \left\langle f \Big| \frac{1}{i} g' - \frac{i}{2} ug \right\rangle = \frac{1}{i} \langle f | g' \rangle - \frac{i}{2} \langle f | ug \rangle$$

$$= -\frac{1}{i} \langle f' | g \rangle - \frac{i}{2} \langle uf | g \rangle = \left\langle \frac{1}{i} f' \Big| g \right\rangle + \left\langle \frac{i}{2} uf \Big| g \right\rangle$$

$$= \left\langle \frac{1}{i} f' + \frac{i}{2} uf \Big| g \right\rangle = \langle b^\dagger f | g \rangle,$$

for all $f, g \in D_0$. As a consequence, the adjoint operator b^* to b is an extension of $b^\dagger$. Hence, b^* is densely-defined, and therefore b is closable.

Further, for $f \in D_0$, it follows that

$$b^*bf = \frac{1}{i}(bf)' + \frac{i}{2}ubf$$

$$= \frac{1}{i}\left(\frac{1}{i}f' - \frac{i}{2}uf\right)' + \frac{i}{2}u\left(\frac{1}{i}f' - \frac{i}{2}uf\right)$$

$$= \frac{1}{i}\left(\frac{1}{i}f'' - \frac{i}{2}f - \frac{i}{2}uf'\right) + \frac{i}{2}u\left(\frac{1}{i}f' - \frac{i}{2}uf\right)$$

$$= -f'' - \frac{1}{2}f - \frac{1}{2}uf' + \frac{1}{2}uf' + \frac{u^2}{4}f$$

$$= -f'' + \frac{u^2}{4}f - \frac{1}{2}f,$$

resulting in

$$\hat{H}_0 f = \hbar\omega\left(b^*b + \frac{1}{2}\right)f,$$

for every $f \in D_0$. This observation opens the path for a direct definition of $\hat{H}$, given by the following self-adjoint extension of $\hat{H}_0$,

$$\hat{H} := \hbar\omega\left(\overline{b}^*\overline{b} + \frac{1}{2}\right),$$

where we use that A^*A, which, as usual, is assumed to be maximally defined, is a DSLO for every closed DLO A, see, e.g., Theorem VIII. 32 of Vol. I, [3].

3.4.3 Properties of $\hat{H}$

In a first step, we note that

$$\overline{b}f_0 = 0,$$

where

$$f_0 := \alpha_0 \exp\left(-\frac{u^2}{4}\right) \in D_0$$

and $\alpha_0 \in \mathbb{C}$. Hence

$$\hat{H} f_0 = \frac{\hbar\omega}{2} f_0 \ .$$

Since

$$\left\langle \exp\left(-\frac{u^2}{4}\right) \mid \exp\left(-\frac{u^2}{4}\right)\right\rangle$$
$$= \int_{-\infty}^{\infty} \exp\left(-\frac{1}{2} u^2\right) du = \sqrt{\frac{\pi}{\frac{1}{2}}} = \sqrt{2\pi},$$

in the following, we define

$$\alpha_0 := \frac{1}{(2\pi)^{1/4}},$$

such that

$$\| f_0 \| = 1 \ .$$

Further, for $f \in D_0$, it follows that

$$b\, b^* f = \frac{1}{i} (b^* f)' - \frac{i}{2} u b^* f$$
$$= \frac{1}{i} \left(\frac{1}{i} f' + \frac{i}{2} uf\right)' - \frac{i}{2} u \left(\frac{1}{i} f' + \frac{i}{2} uf\right)$$
$$= \frac{1}{i} \left(\frac{1}{i} f'' + \frac{i}{2} f + \frac{i}{2} uf'\right) - \frac{i}{2} u \left(\frac{1}{i} f' + \frac{i}{2} uf\right)$$
$$= -f'' + \frac{1}{2} f + \frac{1}{2} uf' - \frac{1}{2} u f' + \frac{u^2}{4} f$$
$$= -f'' + \frac{u^2}{4} f + \frac{1}{2} f = b^* b f + f,$$

i.e., it follows that

$$(b\, b^* - b^* b) f = f,$$

for every $f \in D_0$.

Also, if $\lambda \in \mathbb{R}$ is an eigenvalue of $\hat{H}$ and $f_\lambda \in D_0 \backslash \{0\}$ is a corresponding eigenvector, i.e., such that

$$\hat{H} f_\lambda = \hbar\omega \left(b^* b + \frac{1}{2}\right) f_\lambda = \lambda f_\lambda,$$

then

$$\hat{H}\overline{b}^* f_\lambda = \hbar\omega \left(b^*b + \frac{1}{2}\right) b^* f_\lambda = \hbar\omega \left(b^*b\, b^* f_\lambda + \frac{1}{2} b^* f_\lambda\right)$$

$$= \hbar\omega \left[b^*(f_\lambda + b^*bf_\lambda) + \frac{1}{2} b^* f_\lambda\right]$$

$$= \hbar\omega \left(b^*b^*bf_\lambda + \frac{3}{2} b^* f_\lambda\right)$$

$$= \hbar\omega \left[b^*\left(\frac{\lambda}{\hbar\omega} - \frac{1}{2}\right) f_\lambda + \frac{3}{2} b^* f_\lambda\right]$$

$$= \hbar\omega \left(\frac{\lambda}{\hbar\omega} b^* f_\lambda + b^* f_\lambda\right) = (\lambda + \hbar\omega)\, \overline{b}^* f_\lambda,$$

and

$$\left\langle \overline{b}^* f_\lambda \,\middle|\, \overline{b}^* f_\lambda \right\rangle = \left\langle b^* f_\lambda \,\middle|\, b^* f_\lambda \right\rangle = \left\langle f_\lambda \,\middle|\, b\, b^* f_\lambda \right\rangle = \left\langle f_\lambda \,\middle|\, f_\lambda + b^* bf_\lambda \right\rangle$$

$$= \|f_\lambda\|^2 + \left\langle f_\lambda \,\middle|\, \left(\frac{\lambda}{\hbar\omega} - \frac{1}{2}\right) f_\lambda \right\rangle = \|f_\lambda\|^2 + \|f_\lambda\|^2 \left(\frac{\lambda}{\hbar\omega} - \frac{1}{2}\right)$$

$$= \left(\frac{\lambda}{\hbar\omega} + \frac{1}{2}\right) \|f_\lambda\|^2,$$

i.e., the application of the operator $\overline{b}^*$ to an eigenvector leads to an eigenvector corresponding to a higher eigenvalue. Therefore, *the operator $\overline{b}^*$ is called a raising operator.*

Similarly, the application of the operator $\overline{b}$ to an eigenvector leads to an eigenvector corresponding to a lower eigenvalue. Therefore, *the operator $\overline{b}$ is called a lowering operator.* Also, *the operators $\overline{b}^*$ and $\overline{b}$ are called "ladder operators."* That $\overline{b}$ is a lowering operator can be seen as follows.

If again, $\lambda \in \mathbb{R}$ is an eigenvalue of $\hat{H}$ and $f_\lambda \in D_0 \setminus \{0\}$ is a corresponding eigenvector, i.e., such that

$$\hat{H} f_\lambda = \hbar\omega \left(b^*b + \frac{1}{2}\right) f_\lambda = \lambda f_\lambda;$$

then

$$\hat{H}\overline{b} f_\lambda = \hbar\omega \left(b^*b + \frac{1}{2}\right) bf_\lambda = \hbar\omega \left(b\, b^* - \frac{1}{2}\right) bf_\lambda$$

$$= \hbar\omega \left(b\, b^*bf_\lambda - \frac{1}{2} bf_\lambda\right)$$

$$= \hbar\omega \left[b\left(\frac{\lambda}{\hbar\omega} - \frac{1}{2}\right) f_\lambda - \frac{1}{2} bf_\lambda\right]$$

$$= \hbar\omega \left[\left(\frac{\lambda}{\hbar\omega} - 1\right) bf_\lambda\right]$$

$$= (\lambda - \hbar\omega)\, \overline{b} f_\lambda,$$

and

$$\langle \bar{b} f_\lambda | \bar{b} f_\lambda \rangle = \langle b f_\lambda | b f_\lambda \rangle = \langle f_\lambda | b^* b f_\lambda \rangle$$
$$= \left\langle f_\lambda \left| \left(\frac{\lambda}{\hbar \omega} - \frac{1}{2} \right) f_\lambda \right\rangle = \left(\frac{\lambda}{\hbar \omega} - \frac{1}{2} \right) \| f_\lambda \|^2 \; .$$

In the following, we are going to prove by induction that

$$\hat{H} \bar{b}^{*n} f_0 = \left(n + \frac{1}{2} \right) \hbar \omega \, \bar{b}^{*n} f_0, \quad \| \bar{b}^{*n} f_0 \| = \sqrt{n!}, \tag{3.3}$$

for every $n \in \mathbb{N}^*$. Indeed, if $n = 1$, then

$$\hat{H} \bar{b}^* f_0 = \left(\frac{\hbar \omega}{2} + \hbar \omega \right) \bar{b}^* f_0 = \frac{3 \hbar \omega}{2} \, \bar{b}^* f_0,$$

and

$$\left\langle \bar{b}^* f_0 | \bar{b}^* f_0 \right\rangle = 1 \; .$$

If (3.3) is true for $n \in \mathbb{N}^*$, then

$$\hat{H} \bar{b}^{*(n+1)} f_0 = \hat{H} \bar{b}^* \bar{b}^{*n} f_0 = \left[\left(n + \frac{1}{2} \right) \hbar \omega + \hbar \omega \right] \bar{b}^{*(n+1)} f_0$$
$$= \left(n + 1 + \frac{1}{2} \right) \hbar \omega \, \bar{b}^{*(n+1)} f_0$$

as well as

$$\| \bar{b}^{*(n+1)} f_0 \|^2 = \left\langle \bar{b}^{*(n+1)} f_0 | \bar{b}^{*(n+1)} f_0 \right\rangle$$
$$= \left(\frac{(n + \frac{1}{2}) \hbar \omega}{\hbar \omega} + \frac{1}{2} \right) \| \bar{b}^{*n} f_0 \|^2$$
$$= (n + 1) \| \bar{b}^{*n} f_0 \|^2 = (n + 1) n! = (n + 1)!,$$

implying that (3.3) is true for $n + 1 \in \mathbb{N}^*$. As a consequence,

$$(f_n)_{n \in \mathbb{N}},$$

where

$$f_n := \sqrt{\frac{1}{n! \sqrt{2\pi}}} \, \bar{b}^{*n} \exp \left(-\frac{u^2}{4} \right),$$

for every $n \in \mathbb{N}$, *is an orthonormal sequence of eigenvectors of* $\hat{H}$, where

$$\bar{b}^{*0} \exp \left(-\frac{u^2}{4} \right) := \exp \left(-\frac{u^2}{4} \right),$$

corresponding to the sequence of eigenvalues

$$\left(\left(n + \frac{1}{2} \right) \hbar\omega \right)_{n \in \mathbb{N}},$$

i.e.,

$$\hat{H} f_n = \hbar\omega \left(n + \frac{1}{2} \right) f_n,$$

for every $n \in \mathbb{N}$. In the next step, we are going to show inductively that

$$\bar{b}^{*n} \exp\left(-\frac{u^2}{4} \right) = \frac{1}{i^n} \exp\left(\frac{u^2}{4} \right) \left[\exp\left(-\frac{u^2}{4} \right) \right]^{(n)}, \tag{3.4}$$

for every $n \in \mathbb{N}$. Indeed,

$$\bar{b}^{*0} \exp\left(-\frac{u^2}{4} \right) = \exp\left(-\frac{u^2}{4} \right)$$

$$= \frac{1}{i^0} \exp\left(\frac{u^2}{4} \right) \left[\exp\left(-\frac{u^2}{4} \right) \right]^{(0)}.$$

Further, if (3.4) is true for $n \in \mathbb{N}$, then

$$\bar{b}^{*(n+1)} \exp\left(-\frac{u^2}{4} \right) = \frac{1}{i^n} \bar{b}^* \exp\left(\frac{u^2}{4} \right) \left[\exp\left(-\frac{u^2}{4} \right) \right]^{(n)}$$

$$= \frac{1}{i^n} \frac{1}{i} \exp\left(\frac{u^2}{4} \right)$$

$$\cdot \left[\exp\left(-\frac{u^2}{4} \right) \exp\left(\frac{u^2}{4} \right) \left[\exp\left(-\frac{u^2}{4} \right) \right]^{(n)} \right]'$$

$$= \frac{1}{i^{n+1}} \exp\left(\frac{u^2}{4} \right) \left[\exp\left(-\frac{u^2}{4} \right) \right]^{(n+1)},$$

i.e., (3.4) is true for $n + 1 \in \mathbb{N}$.

We note that

$$\bar{b}^{*n} \exp\left(-\frac{u^2}{4} \right)$$

$$= \frac{1}{i^n} \exp\left(\frac{u^2}{4} \right) \left[\exp\left(-\frac{u^2}{4} \right) \right]^{(n)} = \left(\frac{i}{\sqrt{2}} \right)^n H_n\left(\frac{u}{\sqrt{2}} \right),$$

where, for every $n \in \mathbb{N}$, H_n denotes the *corresponding Hermite polynomial*, as defined in [12] on page 344. Indeed,

$$
\frac{1}{i^n} \exp\left(\frac{u^2}{4}\right) \left[\exp\left(-\frac{u^2}{4}\right)\right]^{(n)}
$$

$$
= \frac{1}{i^n} \exp\left(\frac{(u/\sqrt{2})^2}{2}\right) \left[\exp\left(-\frac{(u/\sqrt{2})^2}{2}\right)\right]^{(n)}
$$

$$
= \frac{(-1)^n}{i^n} \left(\frac{1}{\sqrt{2}}\right)^n (-1)^n \exp\left(\frac{y^2}{2}\right) \frac{d^n}{dy^n} \left[\exp\left(-\frac{y^2}{2}\right)\right]^{(n)}
$$

$$
= \frac{(-1)^n}{(i\sqrt{2})^n} H_n(y) = \frac{(-1)^n}{(i\sqrt{2})^n} H_n\left(\frac{u}{\sqrt{2}}\right) = \left(\frac{i}{\sqrt{2}}\right)^n H_n\left(\frac{u}{\sqrt{2}}\right),
$$

where

$$
y = \frac{u}{\sqrt{2}} .
$$

As a consequence,

$$
f_n = i^n \sqrt{\frac{1}{n!\, 2^n \sqrt{2\pi}}}\, H_n\left(\frac{u}{\sqrt{2}}\right) \exp\left(-\frac{u^2}{4}\right),
$$

for every $n \in \mathbb{N}$. We note that the sequence $h_0, h_1, \ldots$ of *Hermite functions* is defined by

$$
h_n(y) := H_n(y) \exp\left(-\frac{y^2}{2}\right),
$$

for every $n \in \mathbb{N}$ and $y \in \mathbb{R}$. Hence

$$
f_n = i^n \sqrt{\frac{1}{n!\, 2^n \sqrt{2\pi}}}\, h_n\left(\frac{u}{\sqrt{2}}\right),
$$

for every $n \in \mathbb{N}$.

Further, $f_0, f_1, \ldots$ is a Hilbert basis for X.

This can be proved as follows. Since,

$$\exp(2xz - z^2) = \sum_{n=0}^{\infty} \frac{1}{n!} H_n(x) z^n,$$

for every $x \in \mathbb{R}$ and $z \in \mathbb{C}$, see 22.9.17 in [12], it follows that

$$\exp\left(-\frac{u^2}{4}\right) \exp(\sqrt{2}\, u\, y - y^2) = \sum_{n=0}^{\infty} \frac{1}{n!} H_n\left(\frac{u}{\sqrt{2}}\right) \exp\left(-\frac{u^2}{4}\right) y^n$$

$$= \sum_{n=0}^{\infty} \frac{1}{n!}\, \frac{y^n}{i^n \sqrt{\frac{1}{n!\, 2^n \sqrt{2\pi}}}}\, f_n(u) = (2\pi)^{1/4} \sum_{n=0}^{\infty} \frac{(-i\sqrt{2}\, y)^n}{\sqrt{n!}}\, f_n(u),$$

for every $u \in \mathbb{R}$ and $y \in \mathbb{R}$. We note that, for $y \in \mathbb{R}$

$$\left(\frac{(2|y|^2)^n}{n!}\right)_{n \in \mathbb{N}}$$

is summable, with sum $e^{2|y|^2}$, and hence that

$$\left(\frac{(-i\sqrt{2}\, y)^n}{\sqrt{n!}}\, f_n\right)_{n \in \mathbb{N}}$$

is summable in X. As a consequence, for $y \in \mathbb{R}$,

$$\exp\left(-\frac{u^2}{4}\right) \exp(\sqrt{2}\, u\, y - y^2) = (2\pi)^{1/4} \sum_{n=0}^{\infty} \frac{(-i\sqrt{2}\, y)^n}{\sqrt{n!}}\, f_n(u),$$

where u denotes the identical function on $\mathbb{R}$. Hence it follows for $f \in X$ and $y \in \mathbb{R}$ that

$$\int_{-\infty}^{\infty} f^*(u) \exp\left(-\frac{u^2}{4}\right) \exp(\sqrt{2}\, u\, y - y^2)\, du$$

$$= (2\pi)^{1/4} \sum_{n=0}^{\infty} \frac{(-i\sqrt{2}\, y)^n}{\sqrt{n!}}\, \langle f(u) | f_n(u) \rangle\, du.$$

In particular, if f is orthogonal to every f_n, $n \in \mathbb{N}$, then

$$\int_{-\infty}^{\infty} f^*(u) \exp\left(-\frac{u^2}{4} + \frac{y}{2}\, u\right) du = 0,$$

for every $y \in \mathbb{R}$. Since,

$$\left[f^* * \exp\left(-\frac{u^2}{4}\right) \right](y) = \int_{-\infty}^{\infty} f^*(u) \cdot \exp\left(-\frac{(y-u)^2}{4}\right) du$$

$$= \exp\left(-\frac{y^2}{4}\right) \int_{-\infty}^{\infty} f^*(u) \cdot \exp\left(-\frac{u^2}{4} + \frac{y}{2}\, u\right) du$$

for every $y \in \mathbb{R}$, this implies that

$$f^* * \exp\left(-\frac{u^2}{4}\right) = 0 \ .$$

Since

$$f^* * \exp\left(-\frac{u^2}{4}\right) = F_1[(F^{-1}f^*)F^{-1}\exp(-u^2/4)]$$
$$= \sqrt{2}F_1[(F^{-1}f^*)\exp(-v^2)],$$

where $*$ denotes the convolution product, and the Fourier transform $F_1 : L_{\mathbb{C}}^1(\mathbb{R}) \to C_\infty(\mathbb{R}, \mathbb{C})$ is defined in the Appendix, v denotes the identical function on $\mathbb{R}$, and since F_1 is injective, we conclude that $F^{-1}f^*$ vanishes and hence also that f vanishes.

Summarizing the previous results, we proved that if $f \in X$ is such that

$$\langle f | f_n \rangle = 0,$$

for every $n \in \mathbb{N}$, then f vanishes. Therefore, $f_0, f_1, \dots$ is a Hilbert basis for X.[16] As a consequence, the spectrum $\sigma(\hat{H})$ of $\hat{H}$ is given by

$$\sigma(\hat{H}) = \left\{ \hbar\omega\left(n + \frac{1}{2}\right) : n \in \mathbb{N} \right\},$$

and is purely discrete, i.e., the spectrum itself is a discrete subset of the real numbers and consists of eigenvalues of finite multiplicity. In this special case, the eigenvalues are all simple.

This spectrum gives the possible outcomes of measurements of the momentum of the particle in quantum theory. Differently to the classical system, these values are discrete.

We note that for a mass of $m = 0.4$ kg and spring constant $k = 18.33$ N/m for our "macroscopic" classical system, the difference of the energy levels amounts to

$$7.13883 * 10^{-34} \text{Joule} = 4.45571 * 10^{-15} \text{eV},$$

where eV denotes the unit of electron volt, which is too small to be measurable. The discreteness of energy levels is only measurable for microscopic systems, such as atoms. For the hydrogen atom, the difference between energy levels is of the order 1 eV. Such energies are measurable.

---

[16] As consequence, also $f_0, (-i)^1 f_1, (-i)^2 f_2, \dots$ is a Hilbert basis for X consisting of eigenvectors of $\hat{H}$. These eigenvectors are real functions, and, for this reason, might be considered more comfortable to use. On the other hand, as said before, these functions are not observable. For this reason, we do not graph these functions.

3.5   Measurement in General

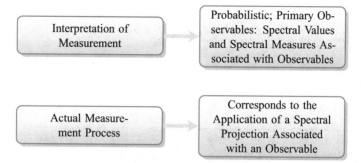

The measurement of an observable A leads without exception to a value from the spectrum $\sigma(A)$ of A. The latter consists of all those complex numbers a for which the corresponding operator $A - a$ is not bijective, i.e., one-to-one onto. Since A is self-adjoint, those values are necessarily *real*.

Measurements follow the rules of probability. The non-normalized *probability* of finding the measured value to be part of a bounded subset $B \subset \mathbb{R}$ is given by[a]

$$\psi_f(B) = \int_{\mathbb{R}} \chi_B \, d\psi_f = \langle f | \chi_B(A) f \rangle \ .$$

Here $\mathbb{C}^* . f$, $f \in X \setminus \{0\}$, is the state of the system, ψ_f is the *spectral measure*[b] associated with A and f, χ_B denotes the characteristic function of B, defined by

$$\chi_B(\lambda) := \begin{cases} 1 & \text{if } \lambda \in B \\ 0 & \text{if } \lambda \notin B \end{cases},$$

for every $\lambda \in \mathbb{R}$, and $\chi_B(A)$ is the *orthogonal projection*[c] which is associated with $\chi_B|_{\sigma(A)}$ by the functional calculus for A.[d]

---

[a] In addition, B is assumed to be a countable union of bounded intervals of $\mathbb{R}$.

[b] ψ_f is an additive, monotone and regular interval function defined on the set of all bounded sub-intervals of $\mathbb{R}$. The existence of these spectral measures is part of the statement of a spectral theorem.

[c] I.e., $P := \chi_B(A)$ is a linear, bounded and self-adjoint operator on X satisfying $P \circ P = P$.

[d] See, e.g., [3] Vol. I, Theorem VIII. 5.

Non spectral values of A do not contribute to this probability, since $\mathbb{R} \setminus \sigma(A)$ is a ψ_f-zero set.[17] In particular, note for the normalization that

$$\psi_f(\mathbb{R}) = \psi_f(\sigma(A)) = \int_{\mathbb{R}} d\psi_f = \langle f | f \rangle \ .$$

Hence the *probability* of finding the measured value to be part of a bounded subset $B \subset \mathbb{R}$ is given by

$$\frac{\psi_f(B)}{\psi_f(\mathbb{R})} = \frac{\int_{\mathbb{R}} \chi_B \, d\psi_f}{\int_{\mathbb{R}} d\psi_f} = \frac{\langle f | \chi_B(A) f \rangle}{\langle f | f \rangle} \ .$$

If a measurement of an observable A finds its value to be part B of its spectrum, the system is in the state

$$\mathbb{C}^* . \chi_B(A) f \ .$$

Note that in the particular case that B is the disjoint union of two bounded subsets $B_1, B_2 \subset \mathbb{R}$ [18] that

$$\mathbb{C}^* . \chi_B(A) f = \mathbb{C}^* . \left(\chi_{B_1}(A) f + \chi_{B_2}(A) f \right),$$

i.e., that the system is in the state corresponding to the *superposition* of $\chi_{B_1}(A) f$ and $\chi_{B_2}(A) f$.

If A is not found to be part of B, it is in the state

$$\mathbb{C}^* . \left(\mathrm{id}_X - \chi_B(A) \right) f = \mathbb{C}^* . \chi_{\mathbb{R} \setminus B}(A) f \ .$$

---

[17] This is the case for all $f \in X$.

[18] For instance this has application in the double-slit experiment.

More generally, according to the usual rules of probability[a], for any bounded function $\eta : \sigma(A) \to \mathbb{R}$[b] being "universally measurable"[c] the non-normalized *expectation value* for the measurement of the observable $\eta(A)$ is given by

$$\int_{\mathbb{R}} \eta \, d\psi_f = \langle f | \eta(A) f \rangle \ .$$

---

[a] $\sigma(A)$ is the sample space and ψ_f is the probability distribution for the random variable $\mathrm{id}_{\sigma(A)}$.
[b] η is a random variable.
[c] In particular, point-wise limits of sequences of continuous functions on $\sigma(A)$ are universally measurable.

Further, as a consequence of the spectral theorems for DSLO's, for every $f \in D(A)$, we have the following relations for the first 2 moments of the spectral measure ψ_f:

$$\int_{\mathbb{R}} \lambda \, d\psi_f = \langle f | Af \rangle, \quad \int_{\mathbb{R}} \lambda^2 \, d\psi_f = \| Af \|^2,$$

where here and in the following λ denotes the identical function on the real numbers. Hence if, in addition, $\| f \| = 1$, the expectation value $\mu_{A,f}$ for the measurement of the observable A is given by

$$\mu_{A,f} = \int_{\mathbb{R}} \lambda \, d\psi_f = \langle f | Af \rangle,$$

and the corresponding variance $\sigma_{A,f}^2$ is given by

$$\sigma_{A,f}^2 = \int_{\mathbb{R}} \left(\lambda - \mu_{A,f} \right)^2 d\psi_f = \int_{\mathbb{R}} \lambda^2 \, d\psi_f - 2\mu_{A,f} \int_{\mathbb{R}} \lambda \, d\psi_f + \mu_{A,f}^2 \int_{\mathbb{R}} d\psi_f$$
$$= \| Af \|^2 - 2\mu_{A,f} \langle f | Af \rangle + \mu_{A,f}^2 \| f \|^2 = \| (A - \mu_{A,f}) f \|^2$$
$$= \| Af \|^2 - 2\mu_{A,f}^2 + \mu_{A,f}^2 = \| Af \|^2 - \mu_{A,f}^2,$$

leading to the standard deviation:

$$\sigma_{A,f} = \| (A - \mu_{A,f}) f \| = \sqrt{\| Af \|^2 - \mu_{A,f}^2} \ .$$

In the following, we are going to use the previous for the derivation of *Heisenberg's uncertainty relations* that relates the standard deviations of the measurement of the 2 observables A and B.

Theorem 3.51 (Heisenberg's Uncertainty Relations) *Let $(X, \langle | \rangle)$ be a non-trivial complex Hilbert space, $A : D(A) \to X$ and $B : D(B) \to X$ be densely-defined, linear and self-adjoint operators in X, without eigenvalues, D a subspace of*

$$\{f \in D(A) \cap D(B) : Af \in D(B) \text{ and } Bf \in D(A)\}$$

and $M : D \to X$ a Hermitian linear operator such that

$$(AB - BA)f = iMf$$

for every $f \in D$. Then

$$\sigma_{A,f} \cdot \sigma_{B,f} \geqslant \frac{1}{2} |\langle f | Mf \rangle|,$$

for every $f \in D$ such that $\|f\| = 1$.

Proof For this purpose, let $f \in D$ such $\|f\| = 1$. Further, let $\alpha \in \mathbb{R}$. Then

$$0 \leqslant \| \left[\alpha.(A - \mu_{A,f}) - i.(B - \mu_{B,f}) \right] f \|^2$$

$$= \langle \alpha(Af - \mu_{A,f}f) - i(Bf - \mu_{B,f}f) | \alpha(Af - \mu_{A,f}f) - i(Bf - \mu_{B,f}f) \rangle$$

$$= \alpha^2 \sigma_{A,f}^2 + \sigma_{B,f}^2$$

$$\quad + i\alpha \left[\langle Bf - \mu_{B,f}f | Af - \mu_{A,f}f \rangle - \langle Af - \mu_{A,f}f | Bf - \mu_{B,f}f \rangle \right]$$

$$= \alpha^2 \sigma_{A,f}^2 + \sigma_{B,f}^2$$

$$\quad + i\alpha \left[\langle Bf | Af \rangle - \mu_{B,f} \langle f | Af \rangle - \mu_{A,f} \langle Bf | f \rangle + \mu_{B,f}\mu_{A,f} \|f\|^2 \right]$$

$$\quad - i\alpha \left[\langle Af | Bf \rangle - \mu_{A,f} \langle f | Bf \rangle - \mu_{B,f} \langle Af | f \rangle + \mu_{A,f}\mu_{B,f} \|f\|^2 \right]$$

$$= \alpha^2 \sigma_{A,f}^2 + \sigma_{B,f}^2$$

$$\quad + i\alpha \left[\langle f | BAf \rangle - \mu_{B,f} \langle f | Af \rangle - \mu_{A,f} \langle f | Bf \rangle + \mu_{A,f}\mu_{B,f} \|f\|^2 \right]$$

$$\quad - i\alpha \left[\langle f | ABf \rangle - \mu_{A,f} \langle f | Bf \rangle - \mu_{B,f} \langle f | Af \rangle + \mu_{A,f}\mu_{B,f} \|f\|^2 \right]$$

$$= \alpha^2 \sigma_{A,f}^2 + \sigma_{B,f}^2 - i\alpha \langle f|(AB - BA)f\rangle$$

$$= \alpha^2 \sigma_{A,f}^2 + \sigma_{B,f}^2 - i\alpha \langle f|iMf\rangle = \alpha^2 \sigma_{A,f}^2 + \sigma_{B,f}^2 + \alpha \langle f|Mf\rangle \ .$$

We not that, since A and B have no eigenvalues, $\sigma_{A,f}$ and $\sigma_{B,f}$ are both different from 0. Hence if

$$\alpha = -\frac{\langle f|Mf\rangle}{2\sigma_{A,f}^2},$$

we arrive at

$$0 \leqslant \frac{(\langle f|Mf\rangle)^2}{4\sigma_{A,f}^4} \cdot \sigma_{A,f}^2 + \sigma_{B,f}^2 - \frac{(\langle f|Mf\rangle)^2}{2\sigma_{A,f}^2} = \sigma_{B,f}^2 - \frac{(\langle f|Mf\rangle)^2}{4\sigma_{A,f}^2}$$

and hence at

$$\sigma_{A,f}^2 \cdot \sigma_{B,f}^2 \geqslant \frac{1}{4} \langle f|Mf\rangle^2 \ .$$

$\square$

3.6 Spectral Measures and Functional Calculus Associated with $\hat{q}$, $\hat{p}$ and $\hat{H}$, Respectively

3.6.1 Position Measurement in Quantum Mechanics in One Space Dimension

We consider the operator $\hat{q}$. This operator corresponds to the measurement of the position. *Hence, here we are using a "position representation."*[19]

According to Operator Theory, its spectrum consists of all real numbers and is purely absolutely continuous.[20] Further, for any $f \in X$ the corresponding spectral measure ψ_f is given by

$$\psi_{\hat{q},f}(I) = \int_{\kappa I} |f(u)|^2 \, du$$

for every bounded interval I of $\mathbb{R}$.[21] *If $f \neq 0$, the quantity $\psi_f(I)$ gives the non-normalized probability in a position measurement of finding the position to be in the range I if the particle is in the state $\mathbb{C}^*.f$* (Fig. 3.1). The corresponding normalized probability is given by

---

[19] The defining property of a "position representation," in 1-space dimension is that the position operator is represented as a maximal multiplication operator by a multiple of an identical function. The inclusion of *multiples* of the identical function is necessary for dimensional reasons. "Momentum representations" are defined analogously.

[20] The latter means that every set of Lebesgue measure 0 is also a set of measure 0 for every spectral measure. In particular, as a consequence, $\hat{q}$ has no eigenvalues.

[21] Note that the elements of I have the dimension of a length. Hence, multiplication by κ gives dimensionless quantities.

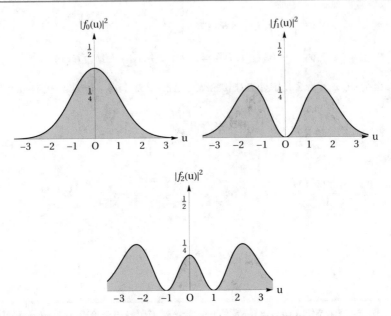

Fig. 3.1 Graphs of the probability distributions associated with the position measurement of f_0, f_1 and f_2. These distributions are observable

$$\frac{\psi_{\hat{q},f}(I)}{\psi_{\hat{q},f}(\mathbb{R})} = \frac{\int_{\kappa I} |f(u)|^2 \, du}{\int_{\mathbb{R}} |f(u)|^2 \, du} \, .$$

More generally, for any bounded function $\eta : \sigma(\hat{q}) \to \mathbb{C}$ being "universally measurable"[22] the corresponding "function" $\eta(\hat{q})$ of $\hat{q}$, a bounded linear operator on X, is given by the maximal multiplication operator defined by

$$\eta(\hat{q})f = \eta\left(\frac{u}{\kappa}\right) \cdot f,$$

for every $f \in X$, where, u denotes the identical function on the real numbers. In particular,

$$e^{i\sigma\hat{q}} f = e^{i\sigma u/\kappa} \cdot f, \tag{3.5}$$

for every $f \in X$, where $\sigma \in \mathbb{R}$ has the dimension of $1/$length.

---

[22] In particular, point-wise limits of sequences of continuous functions on $\sigma(A)$ are universally measurable.

3.6.2 Momentum Measurement in Quantum Mechanics in One Space Dimension

We consider the operator $\hat{p}$. In the position representation for quantum mechanics in one space dimension, this operator corresponds to the measurement of the momentum. *According to Operator Theory*, its spectrum consists of all real numbers and is purely absolutely continuous.[23]

The *change from the position representation to the momentum representation* is achieved *through the Fourier transformation* F, which is a linear unitary transformation and hence a Hilbert space isomorphism. The transformation

$$F : X \to X$$

is defined for every rapidly decreasing test function $f \in \mathscr{S}_{\mathbb{C}}(\mathbb{R})$ by

$$(Ff)(v) := \frac{1}{\sqrt{2\pi}} \int_{-\infty}^{\infty} e^{-ivu} f(u) \, du$$

for every $v \in \mathbb{R}$. Under this transformation, the operator $\hat{p}$ turns into the operator $T_{\hbar\kappa v}$ that multiplies every element of its domain by the function $\hbar\kappa v$, where v denotes the identical function on $\mathbb{R}$, i.e.,

$$D(T_{\hbar\kappa v}) := \{ f \in X : v \cdot f \in X \},$$

and

$$T_{\hbar\kappa v} f := \hbar\kappa v \cdot f,$$

for every $f \in D(T_{\hbar\kappa v})$. We have

$$F\hat{p}F^{-1} = T_{\hbar\kappa v},$$

---

[23] The latter means that every set of Lebesgue measure 0 is also a set of measure 0 for every spectral measure. In particular, as a consequence, $\hat{p}$ has no eigenvalues.

and hence

$$\hat{p} = F^{-1} T_{\hbar \kappa v} F .$$

As a consequence, for any $f \in X$ the corresponding spectral measure $\psi_{\hat{p}, f}$ is given by

$$\psi_{\hat{p}, f}(I) = \int_{(\hbar \kappa)^{-1} I} |(F f)(v)|^2 \, dv$$

for every bounded interval I of $\mathbb{R}$. If $f \neq 0$, the quantity $\psi_{\hat{p}, f}(I)$ gives the non-normalized probability in a momentum measurement of finding the momentum to be in the range I,[24] if the particle is in the state $\mathbb{C}^* . f$ (Fig. 3.2). The corresponding normalized probability is given by

$$\frac{\psi_{\hat{p}, f}(I)}{\psi_{\hat{p}, f}(\mathbb{R})} = \frac{\int_{(\hbar \kappa)^{-1} I} |F f|^2 \, dv}{\int_{\mathbb{R}} |F f|^2 \, dv} .$$

In particular,

$$(F f_0)(v) = \left(\frac{2}{\pi} \right)^{1/4} \exp(-v^2),$$

$$(F f_1)(v) = 2 \left(\frac{2}{\pi} \right)^{1/4} v \exp(-v^2),$$

$$(F f_2)(v) = \frac{1}{(2\pi)^{1/4}} (4v^2 - 1) \exp(-v^2),$$

for every $v \in \mathbb{R}$, leading to

$$|(F f_0)(v)|^2 = \left(\frac{2}{\pi} \right)^{1/2} \exp(-2v^2),$$

$$|(F f_1)(v)|^2 = 4 \left(\frac{2}{\pi} \right)^{1/2} v^2 \exp(-2v^2),$$

$$|(F f_2)(v)|^2 = \frac{1}{(2\pi)^{1/2}} (4v^2 - 1)^2 \exp(-2v^2),$$

for every $v \in \mathbb{R}$.

More generally, since

$$\hat{p} = F^{-1} T_{\hbar \kappa v} F,$$

[24] Note that the elements of I have the dimension of a Mass $\cdot$ Length/Time. Hence, multiplication by $(\hbar \kappa)^{-1}$ gives dimensionless quantities.

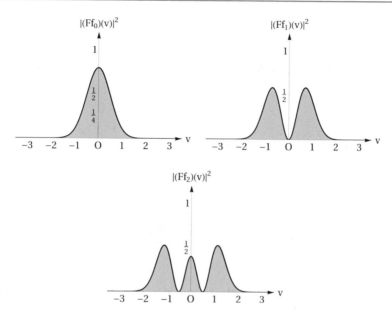

Fig. 3.2 Graphs of the probability distributions associated with the momentum measurement of f_0, f_1 and f_2. These distributions are observable

for any bounded function $\eta : \sigma(\hat{p}) \to \mathbb{C}$ being "universally measurable" [25] the corresponding "function" $\eta(\hat{p})$ of $\hat{p}$, a bounded linear operator on X, is given by the operator defined by

$$\eta(\hat{p}) = F^{-1} T_{\eta(\hbar \kappa v)} F,$$

for every $f \in X$, where, v denotes the identical function on the real numbers.

3.6.3 Weyl's Form of the Canonical Commutation Rules

In particular,

$$e^{it\hat{p}} f = F^{-1} e^{it\hbar\kappa v} F f,$$

for every $f \in X$, where $\tau \in \mathbb{R}$ has the dimension of $1/$Momentum. Since

$$F f(u + a) = e^{iav} F f(u),$$

---

[25] In particular, point-wise limits of sequences of continuous functions on $\sigma(\hat{p})$ are universally measurable.

we have

$$f(u + a) = F^{-1} e^{iau} F f(u),$$

for every $f \in X$ and $a \in \mathbb{R}$, where u denotes the identical function on the real numbers. As a consequence,

$$e^{i\tau \hat{p}} f = F^{-1} e^{i\tau \hbar \kappa v} F f = f(u + \hbar \kappa \tau), \qquad (3.6)$$

for every $f \in X$ and $\tau \in \mathbb{R}$, the latter assumed to have the dimension 1/Momentum.

Using (3.5) and (3.6), *we arrive at Weyl's form of the canonical commutation relations. These relations are required to be satisfied by the quantizations $\hat{p}$ and $\hat{q}$ of the canonically conjugate observables of the classical system, momentum and position.*

$$e^{i\tau \hat{p}} e^{i\sigma \hat{q}} f = e^{i\hbar \tau \sigma} e^{i\sigma \hat{q}} e^{i\tau \hat{p}} f,$$

for every $f \in X$ and real τ, with dimension 1/Momentum, as well as real σ, with dimension 1/length. Another version of Weyl's form of the canonical commutation relations, with dimensionless parameters, is the following

$$e^{i\tau (\hbar \kappa)^{-1} \hat{p}} e^{i\sigma \kappa \hat{q}} f = e^{i\tau \sigma} e^{i\sigma \kappa \hat{q}} e^{i\tau (\hbar \kappa)^{-1} \hat{p}} f,$$

for every $f \in X$ and real τ and σ.

3.6.4   Heisenberg's Uncertainty Relation for $\hat{p}$ and $\hat{q}$

In the following, we are going to apply Theorem 3.51 to the position operator $\hat{q}$ and the momentum operator $\hat{p}$. Both are densely-defined, linear and self-adjoint operators in X, without eigenvalues. In the following, u and v are going to denote the identical functions on the real numbers, in "state space" and in "momentum space," respectively. Since

$$\hat{p} = F^{-1} T_{\hbar \kappa v} F,$$

in a first step, it follows that

$$D_1 := \{ f \in C^1(\mathbb{R}, \mathbb{C}) \cap L^1_{\mathbb{C}}(\mathbb{R}) \cap X : f' \in L^1_{\mathbb{C}}(\mathbb{R}) \cap X \} \subset D(\hat{p})$$

and for $f \in D_1$ that

$$\hat{p}f = \hbar\kappa F^{-1}T_v Ff = \frac{1}{\sqrt{2\pi}}\hbar\kappa F^{-1}(vF_1 f)$$

$$= \frac{1}{\sqrt{2\pi}}\frac{\hbar\kappa}{i}F^{-1}F_1 f' = \frac{\hbar\kappa}{i}.F^{-1}Ff' = \frac{\hbar\kappa}{i}.f' \ .$$

Further, for $f \in D_2$, where

$$D_2 := \{f \in C^1(\mathbb{R}, \mathbb{C}) : f, f', uf, uf' \in L^1_{\mathbb{C}}(\mathbb{R}) \cap X\},$$

it follows that

$$\hat{q}\,\hat{p}f = \frac{\hbar}{i}uf',$$

$$\hat{p}\,\hat{q}f = \frac{\hbar}{i}(uf)' = \frac{\hbar}{i}(f + uf')$$

$$= \frac{\hbar}{i}f + \frac{\hbar}{i}uf' = \frac{\hbar}{i}.f + \hat{q}\,\hat{p}f$$

and hence that $\hat{p}$ and $\hat{q}$ satisfy the canonical commutation rules

$$[\hat{p}, \hat{q}]f = (\hat{p}\,\hat{q} - \hat{q}\,\hat{p})f = \frac{\hbar}{i}f \ .$$

Hence, we conclude from Theorem 3.51 that

$$\sigma_{\hat{q},f} \cdot \sigma_{\hat{p},f} \geqslant \frac{\hbar}{2},$$

for every $f \in D_2$ such that $\|f\|_2 = 1$, implying that the product of the standard deviation of the position measurement and momentum measurement cannot vanish, for the states from the subspace D_2. This result is called Heisenberg's uncertainty relation for the position and the momentum measurement [13].

3.6.5 Energy Measurement for the Harmonic Oscillator

We consider the operator $\hat{H}$ that corresponds to the measurement energy of the harmonic oscillator. Since its spectrum is purely discrete, the spectral measure corresponding to $\hat{H}$ and $f \in X$ are discrete, too. Integration with respect to these measures turns into a summation.

Since, the sequence of eigenstates of $\hat{H}$, $f_0, f_1, \ldots$ form a Hilbert basis of X, for every $f \in X$, we have

$$f = \sum_{n=0}^{\infty} \langle f_n | f \rangle \, f_n,$$

where the convergence of the sum is with respect to the L^2-norm, $\| \, \|_2$, that is induced on X by the scalar product, and

$$\sum_{n=0}^{\infty} |\langle f_n | f \rangle|^2 = \| f \|_2^2 .$$

Hence, if the system is in the state $\mathbb{C}^*.f$, where $f \in X \setminus \{0\}$, the normalized probability of finding the energy of the system to be E_n, where

$$E_n := \hbar\omega\left(n + \frac{1}{2}\right),$$

for every $n \in \mathbb{N}$, is given by

$$\frac{|\langle f_n | f \rangle|^2}{\| f \|^2} .$$

A measurement of the energy of the system forces the system to assume a state $\mathbb{C}^*.f_n$ for one $n \in \mathbb{N}$. If the measured energy of the system is found to be $E_n, n \in \mathbb{N}$, then after the measurement the system is in the state $\mathbb{C}^*.f_n$ and a subsequent energy measurement will give with absolute certainty the same value.

Further, for any bounded function $\eta : \sigma(\hat{H}) \to \mathbb{C}$, the corresponding "function" $\eta(\hat{p})$ of $\hat{H}$, a bounded linear operator on X, is given by

$$\eta(\hat{H})f = \sum_{n=0}^{\infty} \langle f_n | f \rangle \, \eta(E_n) f_n, \tag{3.7}$$

for every $f \in X$.

3.7 Time Evolution of Physical States in General

The so called *Hamilton operator* is associated with the Hamiltonian of the classical system. It is the generator of the time evolution of the states in the following sense.[ab] If the system is in the state $\mathbb{C}^*.f$, $f \in X \setminus \{0\}$, at time $t_0 \in \mathbb{R}$, it will be/was in the state

$$\mathbb{C}^*.U(t - t_0)f$$

at time $t \in \mathbb{R}$. Here

$$U(t) := \exp\left(-i\,(t/\hbar).E\right)(H),$$

where E denotes the identical function on the spectrum $\sigma(H)$ of H, is unitary for every $t \in \mathbb{R}$, and $\hbar$ is the reduced Planck's constant.

---

[a] This is true if the system is closed. Otherwise the Hamiltonian can depend on time.

[b] Here we are using the "*Schrödinger picture*." In the equivalent "*Heisenberg picture*" the observables undergo time-evolution, whereas the states of the system stay the same. Heisenberg's picture is generally used in Quantum Field Theory.

Note that the *unitarity* of time evolution corresponds to *conservation of probability*.

If $f \in D(H)$, the unique solution $u : \mathbb{R} \to D(H)$ of the "*Schrödinger equation*"

$$i\hbar.u'(t) = Hu(t)$$

such that $u(t_0) = f$, where $'$ denotes the ordinary derivative of a X-valued path, is given by

$$u(t) := U(t - t_0)f$$

for all $t \in \mathbb{R}$. [a]

---

[a] See Chapter VIII. 4 in Vol. I of [3] on 'Stone's theorem'. Note that generically $D(H)$ is a proper dense subspace of X. Hence there is no solution to the Schrödinger equation for data from $X \setminus D(H)$.

3.8   Time Evolution of the Physical States for the Harmonic Oscillator

According to (3.7), we have

$$U(t)f = \sum_{n=0}^{\infty} \exp\left(-i\,\frac{E_n t}{\hbar}\right) \langle f_n | f \rangle\, f_n = \sum_{n=0}^{\infty} \exp\left[-i\left(n+\frac{1}{2}\right)\omega t\right] \langle f_n | f \rangle\, f_n,$$

for every $t \in \mathbb{R}$ and $f \in X$. We note that the quantity ωt is dimensionless. In particular, the eigenstates $\mathbb{C}^* . f_n$, $n \in \mathbb{N}$, of the harmonic oscillator are stationary, i.e., do not change with time.

For example, if

$$f = \frac{1}{\sqrt{2}}\,(f_0 - i f_1) = \frac{e^{-\frac{u^2}{4}}(1+u)}{2^{3/4}\pi^{1/4}},$$

then the corresponding probability distribution for the measurement of position is given by

$$\rho = |U(t)f|^2 = \frac{e^{-\frac{u^2}{2}}[1+u^2+2u\cos(\omega t)]}{2\sqrt{2\pi}}, \tag{3.8}$$

for every $t \in \mathbb{R}$. The oscillation that is visible in ρ *indicates the phenomenon of wave interference* (Fig. 3.3).

Fig. 3.3 Graph of the probability distribution ρ, for the measurement of position, as a function of ωt, from (3.8)

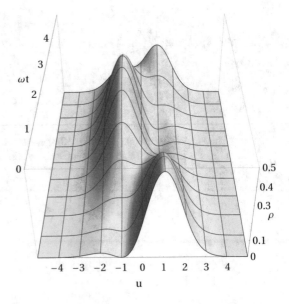

References

1. Born M, Heisenberg W, Jordan P 1926, *Zur Quantenmechanik II*, Zeitschrift für Physik, Vol. **35**, 557–615.
2. Born M, Jordan P 1930, *Elementare Quantenmechanik*, Springer Verlag: Berlin.
3. Reed M and Simon B, 1980, 1975, 1979, 1978, *Methods of modern mathematical physics*, Volume I, II, III, IV, Academic: New York.
4. Riesz F and Sz-Nagy B 1955, *Functional analysis*, Unger: New York.
5. Rudin W 1991, *Functional analysis*, 2nd ed., MacGraw-Hill: New York.
6. Dunford N, Schwartz J T 1957, *Linear operators, Part I: General theory*, Wiley: New York.
7. Goldberg S 1985, *Unbounded linear operators*, Dover: New York.
8. Kato T 1966, Perturbation theory for linear operators, Springer: New York.
9. Schechter M 2003, *Operator methods in quantum mechanics*, Dover Publication: New York.
10. Weidmann J 1980, *Linear Operators in Hilbert spaces*, Springer: New York.
11. Bjorken J D, Drell S D 1964, *Relativistic quantum mechanics*, McGraw-Hill: New York.
12. Abramowitz M and Stegun I A (ed), 1984, *Pocketbook of Mathematical Functions*, Thun: Harri Deutsch.
13. Heisenberg W 1927, *über den anschaulichen Inhalt der quantentheoretischen Kinematik und Mechanik*, Zeitschrift für Physik **43**, 172–198.

Conclusion

4

We conclude with a short summary of the measurement process in quantum theory, in terms of operator theory.

In 1930 and in connection with the development of the mathematical foundation of quantum mechanics, von Neumann proved the spectral theorem for densely-defined, linear and self-adjoint operators in Hilbert spaces [1]. Usually, this theorem comes in 3 forms. Most important for applications is the form that associates a "functional calculus" with every such operator. Adding the obtained insight to the standard ("Copenhagen") picture of quantum mechanics completes the picture. It reveals that, roughly speaking, every such operator is a "physical observable" and the other way around.[1] To every such operator and every element (or "physical state") of the Hilbert space, there corresponds a spectral measure, whose support is part of the spectrum[2] of the operator. These spectral measures are the principal observables of quantum theory; in particular, they can be measured by experiment.[3] The probability of a measurement of the observable A to find the measured value to belong to a

---

[1] A common misconception in standard physics textbooks is to define observables as symmetric (or "Hermitian") linear operators. This condition is significantly weaker than self-adjointness. Now, there are no spectral theorems for the class of symmetric linear operators. Also, there are symmetric linear operators, (like the natural candidate for a momentum operator on the half-line, an operator that is maximally symmetric, but has no self-adjoint extensions), whose spectra include non-real values, thus excluding them as observables, e.g., see Sect. A.3.4.

[2] Not just the eigenvalues.

[3] Every physicist learns that, unlike wave functions themselves, the spectral measures corresponding to the position operator, i.e., the squares of the absolute values of the wave functions, are observable. The same has to be true for other observables. Otherwise, the position operator would be singled out by the theory in an unnatural way. In this connection, it needs to be remembered that position operators are of no relevance in quantum field theory. Hence, the singling out of position operators in the development of quantum theory would be misleading, and actually was misleading in the precursor, "relativistic quantum mechanics", of quantum field theory.

© The Author(s), under exclusive license to Springer Nature Switzerland AG 2022
H. Beyer, *The Reasoning of Quantum Mechanics*, Synthesis Lectures on Engineering, Science, and Technology, https://doi.org/10.1007/978-3-031-17177-2_4

Borel measurable subset of the spectrum[4] is given by the measure of this subset with respect to the spectral measure corresponding to (A and) the physical state, the latter assumed to have norm 1. In particular, the probability of a measurement of finding a value outside the spectrum $\sigma(A)$ of A is 0. If a measurement of A finds the a value inside a Borel measurable set, B, after the measurement, the physical state is given by the image of the state before the measurement under the operator $(\chi_B|_{\sigma(A)})(A)$, where $\chi_B|_{\sigma(A)}$ denotes the restriction of the characteristic function corresponding to B to $\sigma(A)$, corresponding to B, according to the functional calculus that is associated with A. These operators are orthogonal projections. All states in the range of this operator have the property that the probability of finding the value of the observable to be inside the set B is 1, i.e., is absolutely certain.

Reference

1. Von Neumann J 1930, *Allgemeine Eigenwerttheorie Hermitescher Funktionaloperatoren*, Mathematische Annalen, Vol. **102**, 49–131.

---

[4] Which is a non-empty closed subset of the real numbers.

Appendix

A

In the following, we introduce prerequisites for and basics of the language of Operator Theory that can also be found in most textbooks on Functional Analysis [1–5]. For the convenience of the reader, we also include corresponding proofs, but encourage readers of acquiring a more complete picture, e.g., from the above cited sources.

A.1 Normed Vector Spaces

Definition A.1.1 (Normed vector spaces, Banach spaces) Let $\mathbb{K} \in \{\mathbb{R}, \mathbb{C}\}$. A pair $(X, \| \ \|)$

(a) is called a normed vector space over $\mathbb{K}$ if X is a vector space over $\mathbb{K}$ and $\| \ \| : X \to [0, \infty)$ is map such that

  (i) $\|f\| = 0$ if and only if $f = 0_X$ (*i.e.,* $\| \ \|$ *is positive definite*),
  (ii) $\|\lambda f\| = |\lambda| \, \|f\|$ for every $f \in X$ and $\lambda \in \mathbb{K}$ (*i.e.,* $\| \ \|$ *is homogeneous*),
  (iii) $\|f + g\| \leqslant \|f\| + \|g\|$ for all $f, g \in X$ (*i.e.,* $\| \ \|$ *satisfies triangle inequalities*).

(b) is called a Banach space over $\mathbb{K}$ or a $\mathbb{K}$-Banach space if $(X, \| \ \|)$ is a complete normed vector space over $\mathbb{K}$, i.e., if every Cauchy-sequence in X is convergent to an element of X.

Remark A.1.2 If $\mathbb{K} \in \{\mathbb{R}, \mathbb{C}\}$, $(X, \| \ \|)$ and D is a subspace of X, then $(D, \| \ \|_D)$ is a normed vector space, too. Therefore, unless indicated otherwise, we consider every subspace to be automatically equipped with the corresponding restriction of $\| \ \|$.

© The Editor(s) (if applicable) and The Author(s), under exclusive license to Springer Nature Switzerland AG 2022
H. Beyer, *The Reasoning of Quantum Mechanics*, Synthesis Lectures on Engineering, Science, and Technology, https://doi.org/10.1007/978-3-031-17177-2

We define

$$\mathcal{L}_\mathbb{C}^1(\mathbb{R}) := \{f : \mathbb{R} \to \mathbb{C} : \mathrm{Re}(f), \mathrm{Im}(f) \text{ are Lebesgue measurable}$$
$$\text{and } |\mathrm{Re}(f)|, |\mathrm{Im}(f)| \text{ are Lebesgue integrable}\}$$

and for all $f, g \in \mathcal{L}_\mathbb{C}^1(\mathbb{R})$

$$\|f\|_1 := \int_{-\infty}^{\infty} |f(u)| \, du .$$

Then, *according to Functional Analysis,*

$$\boxed{(\mathrm{L}_\mathbb{C}^1(\mathbb{R}), +, . , \| \, \|_1) \text{ is a complex Banach space,}}$$

where

$$\mathrm{L}_\mathbb{C}^1(\mathbb{R}) := \mathcal{L}_\mathbb{C}^1(\mathbb{R})/\sim ,$$

the equivalence relation $\sim$ on $\mathcal{L}_\mathbb{C}^1(\mathbb{R})$ is defined by[1]

$$f \sim g :\Leftrightarrow f = g \text{ a.e. on } \mathbb{R} ,$$

for all $f, g \in \mathcal{L}_\mathbb{C}^1(\mathbb{R})$, and X is *equipped with the operations* $+, .$ *and the scalar product* $\langle \, | \, \rangle$, *defined by*

$$[f] + [g] := [f + g] , \quad \lambda.[f] := [\lambda.f] ,$$
$$\|[f]\|_1 := \|f\|_1 ,$$

for all $f, g \in \mathcal{L}_\mathbb{C}^1(\mathbb{R})$ and $\lambda \in \mathbb{C}$.

As is standard practice, we are not going to indicate that we are working with equivalence classes, rather than functions. Normally, this does not lead to complications, since in applications, usually, the equivalence classes in question, have a unique distinguished, e.g., continuous, representative, which is the basis for considerations. On the other hand, occasionally, it is necessary to remember this fact.

Apart from the unitary Fourier transform F, the text uses a related linear Fourier transformation which is denoted by F_1,

$$\boxed{\mathrm{F}_1 : \mathrm{L}_\mathbb{C}^1(\mathbb{R}) \to \mathrm{C}_\infty(\mathbb{R}, \mathbb{C}) ,}$$

---

[1] a.e. stands for *almost everywhere*, i.e., $\{u \in \mathbb{R} : f(u) \neq g(u)\}$ is set of Lebesgue measure 0.

which for every $f \in L^1_{\mathbb{C}}(\mathbb{R})$, is defined by

$$(F_1 f)(k) := \int_{-\infty}^{\infty} e^{-iku} f(u) \, du$$

for every $k \in \mathbb{R}$. Here, $C_\infty(\mathbb{R}, \mathbb{C})$ denotes the vector space of continuous complex-valued functions on the real numbers that vanish at $\pm\infty$, which, equipped with the norm $\| \, \|_\infty$, defined by

$$\|f\|_\infty := \sup_{k \in \mathbb{R}} |f(k)| \, ,$$

for every $f \in C_\infty(\mathbb{R}, \mathbb{C})$, is a complex Banach space. In particular,

$$\|F_1 f\|_\infty \leqslant \|f\|_\infty \, ,$$

for every $f \in C_\infty(\mathbb{R}, \mathbb{C})$, implying that F_1 is continuous.

A.2 Elementary Properties of Hilbert Spaces

A.2.1 Sesquilinear Forms and Scalar Products

Definition A.2.1 ($\mathbb{K}$-Sesquilinear forms) Let $\mathbb{K} \in \{\mathbb{R}, \mathbb{C}\}$ and X a vector space over $\mathbb{K}$.

(i) s is called a $\mathbb{K}$-Sesquilinear form on X if s is a map from $X \times X$ to $\mathbb{K}$ such that $s(f, \cdot)$ is linear for every $f \in X$ and $s(\cdot, g)$ is linear and anti-linear[2] for every $g \in X$ if $\mathbb{K} = \mathbb{R}$ and $\mathbb{K} = \mathbb{C}$, respectively.

(ii) If s is a $\mathbb{K}$-Sesquilinear form on X, we call the function $(X \to \mathbb{K}, f \mapsto s(f, f))$ the quadratic form that is generated by s.

(iii) A $\mathbb{K}$-Sesquilinear form s on X called Hermitian if $s(f, g) = s(g, f)$ and $s(g, f) = (s(f, g))^*$ for all $f, g \in X$ if $\mathbb{K} = \mathbb{R}$ and $\mathbb{K} = \mathbb{C}$, respectively, where $*$ denotes complex conjugation on $\mathbb{C}$.

(iv) s is called a semi-scalar product and scalar product on X if s is a Hermitian $\mathbb{K}$-Sesquilinear form on X such that $s(f, f) \geqslant 0$ for every $f \in X^3$ and $s(f, f) > 0$ for every $f \in X \setminus \{0\}$,[4] respectively.

Remark A.2.2 Note that Definition A.2.1 (iv) uses that the quadratic forms that are associated with $\mathbb{K}$-sesquilinear forms are real-valued.

---

[2] I.e., $s(f_1 + f_2, g) = s(f_1, g) + s(f_2, g)$ and $s(\lambda f, g) = \lambda^* s(f, g)$ for all $f_1, F, f \in X$ and $\lambda \in \mathbb{C}$, where $*$ denotes complex conjugation on $\mathbb{C}$.

[3] A Hermitian $\mathbb{K}$-Sesquilinear form on X with this property is also called positive semi-definite.

[4] A Hermitian $\mathbb{K}$-Sesquilinear form on X with this property is also called positive definite.

Theorem A.2.3 *(Basic properties of $\mathbb{K}$-Sesquilinear forms) Let $\mathbb{K} \in \{\mathbb{R}, \mathbb{C}\}$, X a vector space over $\mathbb{K}$ and s a $\mathbb{K}$-Sesquilinear form on X. Then*

(i) *(Parallelogram law)*

$$s(f+g, f+g) + s(f-g, f-g) = 2[s(f,f) + s(g,g)]$$

for all $f, g \in X$.

(ii) *(Polarization identity for $\mathbb{C}$-Sesquilinear forms) if $\mathbb{K} = \mathbb{C}$,*

$$s(f,g) = \frac{1}{4}[s(f+g, f+g) - s(f-g, f-g) - is(f+ig, f+ig)$$
$$+ is(f-ig, f-ig)]$$

for all $f, g \in X$.

Proof "(i)":

$$s(f+g, f+g) + s(f-g, f-g) = s(f,f) + s(g,f) + s(f,g) + s(g,g)$$
$$+ s(f,f) - s(g,f) - s(f,g) + s(g,g) = 2[s(f,f) + s(g,g)]$$

for all $f, g \in X$.

"(ii)": First it follows that

$$s(f+g, f+g) = s(f,f) + s(g,f) + s(f,g) + s(g,g) \,,$$
$$s(f-g, f-g) = s(f,f) - s(g,f) - s(f,g) + s(g,g)$$

and hence that

$$s(f+g, f+g) - s(f-g, f-g) = 2[s(f,g) + s(g,f)]$$

for all $f, g \in X$. This implies that

$$-i[s(f+ig, f+ig) - s(f-ig, f-ig)] = 2[s(f,g) - s(g,f)]$$

for all $f, g \in X$. By addition of the last two equations and multiplication of the resulting equation by $1/4$, we arrive at the statement.   □

Remark A.2.4 As a consequence of part (ii) of Theorem A.2.3, a $\mathbb{C}$-Sesquilinear form is uniquely determined by its corresponding quadratic form. That the analogous is not true for $\mathbb{R}$-Sesquilinear forms can be seen from the existence of non-trivial skew-symmetric bilinear forms. The quadratic forms corresponding to the latter vanish.

Theorem A.2.5 *Let $\mathbb{K} \in \{\mathbb{R}, \mathbb{C}\}$, X a vector space over $\mathbb{K}$ and $\langle\,|\,\rangle$ a semi-scalar product on X and $\|\ \| : X \to [0, \infty)$ defined by $\|f\| := \langle f|f \rangle^{1/2}$ for every $f \in X$. Then*

(i) (**Cauchy-Schwarz inequality**) *For all $f, g \in X$, the following Cauchy-Schwarz inequality holds*

$$|\langle f|g \rangle| \leqslant \|f\|\|g\| . \tag{A.1}$$

(ii) (**Triangle inequality**) *For all $f, g \in X$*

$$\|f + g\| \leqslant \|f\| + \|g\| . \tag{A.2}$$

(iii) $N := \{f \in X : \|f\| = 0\}$ *is subspace of X, the so called null space of $\langle\,|\,\rangle$.*

(iv) (**Cauchy-Schwarz equality**)

$$|\langle f|g \rangle| = \|f\|\|g\| \tag{A.3}$$

for some $f, g \in X$ if and only if

$$\|f\|^2 g - \langle f|g \rangle f \in N . $$

Proof "(i)": For arbitrary $f, g \in X$, it follows that

$$\begin{aligned} 0 &\leqslant \langle \|f\|^2 g - \langle f|g \rangle f, \|f\|^2 g - \langle f|g \rangle f \rangle \\ &= \|f\|^2(\|f\|^2\|g\|^2 + |\langle f|g \rangle|^2 - |\langle f|g \rangle|^2 - |\langle f|g \rangle|^2) \\ &= \|f\|^2(\|f\|^2\|g\|^2 - |\langle f|g \rangle|^2) \end{aligned} \tag{A.4}$$

and hence also that

$$0 \leqslant \|g\|^2(\|f\|^2\|g\|^2 - |\langle f|g \rangle|^2) . $$

As a consequence, we conclude the validity of (A.1) if $\|f\| \neq 0$ and/or $\|g\| \neq 0$. Further, if $\|f\| = \|g\| = 0$, it follows that

$$\begin{aligned} 0 &\leqslant \left\langle -af + \frac{1}{2}g \,\middle|\, -af + \frac{1}{2}g \right\rangle = -\frac{1}{2}(\langle af|g \rangle + \langle g|af \rangle) = -\mathrm{Re}(a\,\langle g|f \rangle) \\ &= -|\langle f|g \rangle| , \end{aligned}$$

where $a \in \mathbb{K}$ is such that $|a| = 1$ and

$$a\,\langle g|f \rangle = |\langle f|g \rangle| , $$

and hence that

$$\langle f|g \rangle = 0 . $$

As a consequence, we conclude also in this case the validity of (A.1).

"(ii)": With the help of (i), it follows for arbitrary $f, g \in X$ that

$$\|f + g\|^2 = |\langle f + g | f + g \rangle| = |\|f\|^2 + \|g\|^2 + \langle f|g\rangle + \langle g|f\rangle|$$
$$\leqslant \|f\|^2 + \|g\|^2 + 2\|f\|\|g\| = (\|f\| + \|g\|)^2 .$$

and hence (A.2).

"(iii)": First, as a consequence of the linearity of $\langle 0_X|\cdot\rangle$, $\|0_X\| = \langle 0_X|0_X\rangle^{1/2} = 0$ and hence $0_X \in N$. Further, if $f, g \in N$, it follows by (A.2) that $\|f + g\| = 0$ and hence that $f + g \in N$. Finally, for $f \in N$ and $\lambda \in \mathbb{K}$, it follows that

$$\|\lambda f\|^2 = \langle \lambda f|\lambda f\rangle = |\lambda|^2 \|f\|^2 = 0$$

and hence that $\lambda f \in N$.

"(iv)": If (A.3) is valid for some $f, g \in X$, it follows from (A.4) that $\|f\|^2 g - \langle f|g\rangle f \in N$. If $\|f\|^2 g - \langle f|g\rangle f \in N$, it follows from (A.4) that

$$\|f\|^2 (\|f\|^2 \|g\|^2 - |\langle f|g\rangle|^2) = 0 .$$

If $\|f\| = 0$, it follows by (A.1) the validity of (A.3). $\square$

Theorem A.2.6 *Let* $\mathbb{K} \in \{\mathbb{R}, \mathbb{C}\}$, *$X$ vector space over $\mathbb{K}$, $\langle | \rangle$ a scalar product on X and $\| \| : X \to [0, \infty)$ defined by $\|f\| := \langle f|f\rangle^{1/2}$ for every $f \in X$. Then*

(i) *$(X, \| \|)$ is a normed vector space over $\mathbb{K}$. In the following, we call $\| \|$ the norm that is induced on X by $\langle | \rangle$.*

(ii) *The maps $\langle f|\cdot\rangle$ and $\langle\cdot|f\rangle$, interpreted as maps from $(X, \| \|)$ to $(\mathbb{K}, | |)$, are continuous for every $f \in X$. In particular, if X is non-trivial, $\| \langle f|\cdot\rangle \| = \|f\|$ for every $f \in X$.*

Proof "(i)": First, $\|0_X\| = \langle 0_X|0_X\rangle^{1/2} = 0$, as a consequence of the linearity of $\langle 0_X|\cdot\rangle$. Also, since $\|f\| = \langle f|f\rangle^{1/2} > 0$ for every $f \in X \setminus \{0_X\}$, from $\|f\| = 0$ for some $f \in X$, it follows that $f = 0_X$. Second, $\|\lambda f\| = \langle \lambda f|\lambda f\rangle^{1/2} = (|\lambda|^2 \langle f|f\rangle)^{1/2} = |\lambda| \langle f|f\rangle^{1/2} = |\lambda| \|f\|$ for every $f \in X$ and $\lambda \in \mathbb{K}$. Finally, according to Theorem A.2.5 (ii), $\|f + g\| \leqslant \|f\| + \|g\|$ for all $f, g \in X$. Hence $(X, \| \|)$ is a normed vector space over $\mathbb{K}$.

"(ii)": Let $f \in X$. Then, we conclude by help of the Cauchy-Schwarz inequality, Theorem A.2.5 (i), that

$$|\langle f|g_1\rangle - \langle f|g_2\rangle| = |\langle f|g_1 - g_2\rangle| \leqslant \|f\|\|g_1 - g_2\|$$
$$|\langle g_1|f\rangle - \langle g_2|f\rangle| = |\langle g_1 - g_2|f\rangle| \leqslant \|f\|\|g_1 - g_2\|$$

for all $g_1, g_2 \in X$. The latter implies the continuity of $\langle f|\cdot\rangle$ and $\langle\cdot|f\rangle$. Further, if X is non-trivial, the last implies that $\| \langle f|\cdot\rangle \|_{op} \leqslant \|f\|$. In this case, it also follows that $|\langle f|\cdot\rangle (\|f\|^{-1} f)| = |\langle f|\|f\|^{-1} f\rangle| = \|f\|$ and hence that $\| \langle f|\cdot\rangle \|_{op} \geqslant \|f\|$. $\square$

A.2.2 Hilbert Spaces

Definition A.2.7 (Pre-Hilbert spaces and Hilbert spaces) Let $\mathbb{K} \in \{\mathbb{R}, \mathbb{C}\}$. A pair $(X, \langle | \rangle)$ is called a pre-Hilbert space over $\mathbb{K}$ if X is a vector space over $\mathbb{K}$ and $\langle | \rangle : X^2 \to \mathbb{K}$ is a scalar product on X. If moreover, $(X, \| \ \|)$ is complete, where $\| \ \|$ is the norm that is induced on X by $\langle | \rangle$, we call $(X, \langle | \rangle)$ a Hilbert space.

A.3 Linear Operators in Banach and Hilbert Spaces

A.3.1 Linear Operators in Banach Spaces

Lemma A.3.1 (*Direct sums of Banach and Hilbert spaces*)

(i) *Let $(X, \| \ \|_X)$ and $(Y, \| \ \|_Y)$ be Banach spaces over $\mathbb{K} \in \{\mathbb{R}, \mathbb{C}\}$ and $\| \ \|_{X \times Y} : X \times Y \to \mathbb{R}$ be defined by*

$$\|(f, g)\|_{X \times Y} := \sqrt{\|f\|_X^2 + \|g\|_Y^2}$$

for all $(f, g) \in X \times Y$. Then $(X \times Y, \| \ \|_{X \times Y})$ is a Banach space.

(ii) *Let $(X, \langle | \rangle_X)$ and $(Y, \langle | \rangle_Y)$ be Hilbert spaces over $\mathbb{K} \in \{\mathbb{R}, \mathbb{C}\}$ and $\langle | \rangle_{X \times Y} : (X \times Y)^2 \to \mathbb{K}$ be defined by*

$$\langle (f, g) | (h, k) \rangle_{X \times Y} := \langle f | h \rangle_X + \langle g | k \rangle_Y$$

for all $(f, g), (h, k) \in X \times Y$. Then $(X \times Y, \langle | \rangle_{X \times Y})$ is a Hilbert space.

Proof '(i)': Obviously, $\| \ \|_{X \times Y}$ is positive definite and homogeneous. Further, it follows for $(f, g), (h, k) \in X \times Y$ by the Cauchy-Schwarz inequality for the Euclidean scalar product for $\mathbb{R}^2$ that

$$
\begin{aligned}
\|(f, g) + (h, k)\|_{X \times Y}^2 &= \|f + h\|_X^2 + \|g + k\|_Y^2 \\
&\leqslant (\|f\|_X + \|h\|_X)^2 + (\|g\|_Y + \|k\|_Y)^2 = (a + a')^2 + (b + b')^2 \\
&= a^2 + b^2 + a'^2 + b'^2 + 2(a a' + b b') \\
&\leqslant a^2 + b^2 + a'^2 + b'^2 + 2\sqrt{a^2 + b^2} \cdot \sqrt{a'^2 + b'^2} \\
&= \left(\sqrt{a^2 + b^2} + \sqrt{a'^2 + b'^2} \right)^2 = (\|(f, g)\|_{X \times Y} + \|(h, k)\|_{X \times Y})^2 ,
\end{aligned}
$$

where $a := \|f\|_X, a' := \|h\|_X, b := \|g\|_Y, b' := \|k\|_Y$, and hence that

$$\|(f, g) + (h, k)\|_{X \times Y} \leqslant \|(f, g)\|_{X \times Y} + \|(h, k)\|_{X \times Y} .$$

The completeness of $(X \times Y, \| \ \|_{X \times Y})$ is an obvious consequence of the completeness of X and Y.

'(ii)': Obviously, $\langle \ | \ \rangle_{X \times Y}$ is a positive definite symmetric bilinear, positive definite Hermitian sesquilinear form, respectively. Further, the induced norm on $X \times Y$ coincides with the norm defined in (i). $\qquad\qquad\square$

Definition A.3.2 (Linear Operators) Let $(X, \| \ \|_X)$ and $(Y, \| \ \|_Y)$ be Banach spaces over $\mathbb{K} \in \{\mathbb{R}, \mathbb{C}\}$. Then we define

(i) A map A is called a Y-*valued linear operator in* X if its *domain* $D(A)$ is a subspace of X, Ran $A \subset Y$ and A is linear. If $(Y, \| \ \|_Y) = (X, \| \ \|_X)$ such a map is also called a *linear operator in* X.

(ii) If in addition A is a Y-valued linear operator in X:

    (a) The *graph* $G(A)$ of A by

$$G(A) := \{(f, Af) \in X \times Y : f \in D(A)\} \ .$$

    Note that $G(A)$ is a subspace of $X \times Y$.

    (b) A is *densely-defined* if $D(A)$ is in particular dense in X.

    (c) A is *closed* if $G(A)$ is a closed subspace of $(X \times Y, \| \ \|_{X \times Y})$.

    (d) A Y-valued linear operator B in X is said to be an *extension* of A, symbolically denoted by

$$A \subset B \quad \text{or} \quad B \supset A \ ,$$

    if $G(A) \subset G(B)$.

    (e) A is *closable* if there is a closed extension. In this case,

$$\bigcap_{B \supset A, B \text{ closed}} G(B)$$

    is a closed subspace of $X \times Y$ which, obviously, is the graph of a unique Y-valued closed linear extension $\bar{A}$ of A, called the *closure* of A. By definition, every closed extension B of A satisfies $B \supset \bar{A}$.

    (f) If A is closed, a *core* of A is a subspace D of its domain such that the closure of $A|_D$ coincides with A, i.e., if

$$\overline{A|_D} = A \ .$$

Theorem A.3.3 *(Elementary properties of linear operators)* Let $(X, \| \ \|_X), (Y, \| \ \|_Y)$ be *Banach spaces over* $\mathbb{K} \in \{\mathbb{R}, \mathbb{C}\}$, A *a* Y-valued linear operator in X and $B \in L(X, Y)$.

(i) $(D(A), \| \ \|_A)$, where $\| \ \|_A : D(A) \to \mathbb{R}$ is defined by

$$\|f\|_A := \|(f, Af)\|_{X \times Y} = \sqrt{\|f\|_X^2 + \|Af\|_Y^2}$$

for every $f \in D(A)$, is a normed vector space. Further, the inclusion $\iota_A : (D(A), \| \ \|_A) \hookrightarrow X$ is continuous and $A \in L((D(A), \| \ \|_A), Y)$.
(ii) A is closed if and only if $(D(A), \| \ \|_A)$ is complete.
(iii) If A is closable, then $G(\bar{A}) = \overline{G(A)}$.
(iv) (Bounded inverse theorem) If A is closed and bijective, then $A^{-1} \in L(Y, X)$.
(v) (Closed graph theorem) In addition, let $D(A) = X$. Then A is bounded if and only if A is closed.
(vi) If A is closable, then $A + B$ is also closable and

$$\overline{A + B} = \bar{A} + B \ .$$

Proof '(i)': Obviously, $(D(A), \| \ \|_A)$ is a normed vector space. Further, because of

$$\|\iota_A f\|_X = \|f\|_X \leqslant \sqrt{\|f\|_X^2 + \|Af\|_Y^2} = \|f\|_A$$

and

$$\|Af\|_Y \leqslant \sqrt{\|f\|_X^2 + \|Af\|_Y^2} = \|f\|_A$$

for every $f \in D(A)$, it follows that $\iota_A \in L((D(A), \| \ \|_A), X)$ and $A \in L((D(A), \| \ \|_A), Y)$.
'(ii)': Let A be closed and $f_0, f_1, \ldots$ a Cauchy sequence in $(D(A), \| \ \|_A)$. Then (f_0, Af_0), $(f_1, Af_1), \ldots$ is a Cauchy sequence in $G(A)$ and hence by Lemma A.3.1 along with the closedness of $G(A)$ convergent to some $(f, Af) \in G(A)$. This implies that

$$\lim_{\nu \to \infty} \|f_\nu - f\|_A = 0 \ .$$

Let $(D(A), \| \ \|_A)$ be complete and $(f, g) \in \overline{G(A)}$. Then there is a sequence (f_0, Af_0), $(f_1, Af_1), \ldots$ in $G(A)$ which is convergent to (f, g). Hence $(f_0, Af_0), (f_1, Af_1), \ldots$ is a Cauchy sequence in $X \times Y$. As a consequence, $f_0, f_1, \ldots$ is a Cauchy sequence in $(D(A), \| \ \|_A)$ and therefore convergent to some $h \in D(A)$. In particular,

$$\lim_{\nu \to \infty} \|(f_\nu, Af_\nu) - (h, Ah)\|_{X \times Y} = 0$$

and hence $(f, g) = (h, Ah) \in G(A)$.

'(iii)': Let A be closable. Then the closed graph of every closed extension of A contains $G(A)$ and hence also $\overline{G(A)}$. Therefore $G(\tilde{A}) \supset \overline{G(A)}$. This implies in particular that $\overline{G(A)}$ is the graph of a map $\tilde{A}$. Further, $D(\tilde{A}) = \mathrm{pr}_1 \overline{G(A)}$, where $\mathrm{pr}_1 := (X \times Y \to X, (f, g) \mapsto f)$, is a subspace of X and $\tilde{A}$ is in particular a linear closed extension of A. Hence $\tilde{A} \supset \bar{A}$ and $\overline{G(A)} = G(\tilde{A}) \supset G(\bar{A})$.

'(iv)': Let A be closed and bijective. Then it follows by (ii) that $(D(A), \| \ \|_A)$ is a Banach space and that $A \in L((D(A), \| \ \|_A), Y)$. Hence it follows by the 'bounded inverse theorem theorem', for e.g. see Theorem III.11 in Vol. I of [1], that $A^{-1} \in L(Y, (D(A), \| \ \|_A))$ and by the continuity of ι_A that $A^{-1} \in L(Y, X)$.

'(v)': Let $D(A) = X$. If A is bounded and $f_0, f_1, \ldots$ is some Cauchy sequence in $(X, \| \ \|_A)$, it follows by the continuity of ι_A that $f_0, f_1, \ldots$ is a Cauchy sequence in X and hence convergent to some $f \in X$. Since A is continuous, it follows the convergence of $Af_0, Af_1, \ldots$ to Af and therefore also the convergence of $f_0, f_1, \ldots$ in $(X, \| \ \|_A)$ to f. Hence $(X, \| \ \|_A)$ is complete and A is closed by (ii). If A is closed, it follows by (ii) that $(X, \| \ \|_A)$ is a Banach space and that the bijective X-valued linear operator ι_A is continuous. Hence ι_A is closed by the previous part of the proof. Therefore, the inverse of ι_A is continuous by (iv) and hence A is bounded.

'(vi)': Let A be closable. In a first step, we prove that $\bar{A} + B$ is closed. For this, let $(f, g) \in \overline{G(\bar{A} + B)}$. Then there is a sequence $f_0, f_1, \ldots$ in $D(\bar{A})$ which is convergent to f and such that $(\bar{A} + B)f_0, (\bar{A} + B)f_1, \ldots$ is convergent to g. Since B is continuous, it follows that $\bar{A}f_0, \bar{A}f_1, \ldots$ is convergent to $g - Bf$. Since $\bar{A}$ is closed, it follows that $f \in D(\bar{A})$ as well as $\bar{A}f = g - Bf$ and hence that $f \in D(\bar{A} + B)$ as well as $(\bar{A} + B)f = g$. Hence $\bar{A} + B$ is closed, and therefore $A + B$ is closable such that $\bar{A} + B \supset \overline{A + B}$. Further, it follows by the previous part of the proof that $\overline{A + B} - B$ is a closed extension of A. Hence $\overline{A + B} - B \supset \bar{A}$ and therefore also $\overline{A + B} \supset \bar{A} + B$. Finally, it follows that $\overline{A + B} = \bar{A} + B$. $\qquad \square$

A.3.2 Linear Operators in Hilbert Spaces

Theorem A.3.4 (*Definition and elementary properties of the adjoint*) *Let* $(X, \langle \ | \ \rangle_X)$ *and* $(Y, \langle \ | \ \rangle_Y)$ *be Hilbert spaces over* $\mathbb{K} \in \{\mathbb{R}, \mathbb{C}\}$, *$A$ a densely-defined Y-valued linear operator in X and* $U : X \times Y \to Y \times X$ *the Hilbert space isomorphism defined by* $U(f, g) := (-g, f)$ *for all* $(f, g) \in X \times Y$.

(i) *Then the closed subspace*

$$[U(G(A))]^{\perp} = \{(f, h) \in Y \times X : \langle f | Ag \rangle_Y = \langle h | g \rangle_X$$
$$\text{for all } g \in D(A)\}$$

of $Y \times X$ is the graph of an uniquely determined X-valued linear operator A^* in Y which is in particular closed and called the adjoint of A. If in addition $(X, \langle \,|\, \rangle_X) = (Y, \langle \,|\, \rangle_Y)$, we call A symmetric if $A^* \supset A$ and self-adjoint if $A^* = A$.

(ii) If B is a Y-valued linear operator in X such that $B \supset A$, then

$$B^* \subset A^* \,.$$

(iii) If A^* is densely-defined, then $A \subset A^{**} := (A^*)^*$ and hence A is in particular closable.
(iv) If A is closed, then A^* is densely-defined and $A^{**} = A$.
(v) If A is closable, then $\bar{A} = A^{**}$.
(vi) If $B \in L(X, X)$, then $(A + B)^* = A^* + B^*$.

If in addition $(X, \langle \,|\, \rangle_X) = (Y, \langle \,|\, \rangle_Y)$:

(vii) (Maximality of self-adjoint operators) If A is self-adjoint and $B \supset A$ is symmetric, then $B = A$.
(viii) If A is symmetric, then $\bar{A}$ is symmetric, too. Therefore, we call a symmetric A essentially self-adjoint if $\bar{A}$ is self-adjoint.
(ix) (Hellinger-Toeplitz) If $D(A) = X$ and A is self-adjoint, then $A \in L(X, X)$.

Proof '(i)': First, it follows that

$$[U(G(A))]^{\perp} = \{(g, f) \in Y \times X : \langle (g, f) | U(h, Ah) \rangle_{Y \times X} = 0 \text{ for all } h \in D(A)\}$$

and hence that

$$[U(G(A))]^{\perp} = \{(g, f) \in Y \times X : \langle g | Ah \rangle_Y = \langle f | h \rangle_X \text{ for all } h \in D(A)\} \,.$$

In particular, it follows for $(g, f_1), (g, f_2) \in [U(G(A))]^{\perp}$ that

$$\langle f_1 - f_2 | h \rangle_X = 0$$

for all $h \in D(A)$ and hence that $f_1 = f_2$ since $D(A)$ is dense in X. As a consequence, by

$$A^* : \mathrm{pr}_1[U(G(A))]^{\perp} \to X \,,$$

where $\mathrm{pr}_1 := (Y \times X \to Y, (g, f) \mapsto g)$, defined by

$$A^* g := f \,,$$

for all $g \in \mathrm{pr}_1[U(G(A))]^{\perp}$, where $f \in X$ is the unique element such that $(g, f) \in [U(G(A))]^{\perp}$, there is defined a map such that

$$G(A^*) = [U(G(A))]^{\perp} \,.$$

Note that the domain of A^* is a subspace of Y. In particular, it follows for all $g, k \in D(A^*)$ and $\lambda \in \mathbb{K}$

$$\langle g + k | Ah \rangle_Y = \langle g | Ah \rangle_Y + \langle k | Ah \rangle_Y = \langle A^*g | h \rangle_X + \langle A^*k | h \rangle_X$$
$$= \langle A^*g + A^*k | h \rangle_X$$
$$\langle \lambda.g | Ah \rangle_Y = \lambda^{(*)} \cdot \langle g | Ah \rangle_Y = \lambda^{(*)} \cdot \langle A^*g | h \rangle_X = \langle \lambda.A^*g | h \rangle_X$$

for all $h \in D(A)$ and hence also the linearity of A^*.

'(ii)': Since

$$U(G(B)) \supset U(G(A)) ,$$

it follows that

$$G(B^*) = [U(G(B))]^\perp \subset [U(G(A))]^\perp = G(A^*) .$$

'(iii)': For this, let A^* be densely-defined. Then, it follows

$$\big(Y \times X \to X \times Y, (g, f) \mapsto (-f, g)\big) = -U^{-1}$$

and hence

$$G(A^{**}) = \big[-U^{-1}(G(A^*))\big]^\perp = \big[U^{-1}(G(A^*))\big]^\perp = \big[U^{-1}[U(G(A))]^\perp\big]^\perp$$
$$= \big[[U^{-1}U(G(A))]^\perp\big]^\perp = G(A)^{\perp\perp} = \overline{G(A)} \supset G(A) . \tag{A.5}$$

'(iv)': For this, let A be closed. Then, it follows for $g \in [D(A^*)]^\perp$

$$(0, g) \in \big[U^{-1}(G(A^*))\big]^\perp = \big[U^{-1}[U(G(A))]^\perp\big]^\perp = \big[[U^{-1}U(G(A))]^\perp\big]^\perp$$
$$= G(A)^{\perp\perp} = \overline{G(A)} = G(A)$$

and hence $g = 0$. Hence $D(A^*)$ is dense in X, and it follows by (A.5) that $G(A^{**}) = \overline{G(A)} = G(\bar{A}) = G(A)$.

'(v)': For this, let A be closable. Since $\bar{A}$ is densely defined and closed, it follows by (iv) that $\bar{A}^*$ is densely-defined. Because of $A \subset \bar{A}$, this implies that $A^* \supset \bar{A}^*$ and hence that A^* is densely-defined, too. Therefore, it follows by (iii) that $A \subset A^{**}$ and by (A.5) that $G(A^{**}) = \overline{G(A)} = G(\bar{A})$ and hence, finally, that $A^{**} = \bar{A}$.

'(vi)': Note that, by Riesz' representation theorem, $D(B^*) = Y$. '$A^* + B^* \supset (A + B)^*$': If $f \in D((A + B)^*), g \in D(A)$, then

$$\langle (A + B)^* f | g \rangle_X = \langle f | (A + B)g \rangle_Y = \langle f | Ag \rangle_Y + \langle B^* f | g \rangle_X .$$

The latter implies that

$$\langle f | Ag \rangle_Y = \langle (A + B)^* f - B^* f | g \rangle_X$$

and hence that $f \in D(A^*)$ and

$$A^* f = (A + B)^* f - B^* f \; .$$

The latter implies that

$$(A + B)^* f = (A^* + B^*) f \; .$$

'$(A + B)^* \supset A^* + B^*$': If $f \in D(A^*), g \in D(A)$, then

$$\langle f | (A + B) g \rangle_{\mathrm{Y}} = \langle f | A g \rangle_{\mathrm{Y}} + \langle f | B g \rangle_{\mathrm{Y}} = \langle A^* f | g \rangle_{\mathrm{X}} + \langle B^* f | g \rangle_{\mathrm{X}}$$
$$= \langle (A^* + B^*) f | g \rangle_{\mathrm{X}} \; .$$

Hence $f \in D((A + B)^*)$ and

$$(A + B)^* f = (A^* + B^*) f \; .$$

In the following, it is assumed that $(X, \langle | \rangle_{\mathrm{X}}) = (Y, \langle | \rangle_{\mathrm{Y}})$.
'(vii)': For this, let A be self-adjoint and B a symmetric extension of A. Then, it follows by using $G(B) \supset G(A)$ that

$$G(B) \subset G(B^*) = [U(G(B))]^{\perp} \subset [U(G(A))]^{\perp} = G(A^*) = G(A)$$

and hence $B \subset A \subset B$ and therefore, finally, that $B = A$.
'(viii)': For this, let A be symmetric. Then $A^* \supset A$ and hence also $A^* \supset \bar{A}$.

$$G(\bar{A}^*) = [U(G(\bar{A}))]^{\perp} = [U \, \overline{G(A)} \,]^{\perp} = [\, \overline{U(G(A))} \,]^{\perp} = \left[[U(G(A))]^{\perp\perp} \right]^{\perp}$$
$$= \overline{[U(G(A))]^{\perp}} = \overline{G(A^*)} = G(A^*) \supset G(\bar{A}) \; .$$

Hence it follows that $\bar{A}^* \supset \bar{A}$.
'(ix)': For this, let A be self-adjoint and $D(A) = X$. Then, $A = A^*$ is in particular closed and hence by Theorem A.3.3 (v) bounded. $\square$

A.3.3   Basic Criteria for Self-Adjointness

Theorem A.3.5 (*Basic criteria for essential self-adjointness*) *Let* $(X, \langle | \rangle)$ *be a non-trivial complex Hilbert space and* $A : D(A) \to X$ *be a densely-defined, linear and symmetric Operator in* X.

(i) *If* Ran A *is dense in* X *and there is* $a \in (0, \infty)$ *such*

$$\| A f \| \geqslant a \| f \|$$

for every $f \in D(A)$, *then* A *is essentially self-adjoint.*

(ii) If A is semi-bounded from below with lower bound $\gamma \in \mathbb{R}$, i.e.,

$$\langle f | A f \rangle \geqslant \gamma \, \langle f | f \rangle$$

for every $f \in D(A)$, and there is $\gamma' < \gamma$ such that $\mathrm{Ran}(A - \gamma')$ is dense in X, then A is essentially self-adjoint.

Proof '(i)': For this, let Ran A be dense in X and $a \in (0, \infty)$ be such

$$\| A f \| \geqslant a \| f \|$$

for every $f \in D(A)$. First, since $\bar{A}$ is symmetric, it follows that $\bar{A}^* \supset \bar{A}$. In the following, we show that

$$\bar{A}^* \subset \bar{A} \; . \tag{A.6}$$

For this, let $f \in D(\bar{A}^*)$. Since Ran A is dense in X, there is a sequence $f_1, F, \dots$ in $D(A)$ such that

$$\lim_{\nu \to \infty} A f_\nu = \bar{A}^* f .$$

Since

$$\| A f_\mu - A f_\nu \| \geqslant a \| f_\mu - f_\nu \|$$

for all $\nu, \mu \in \mathbb{N}^*$, $f_1, F, \dots$ is a Cauchy sequence in X and hence, by the completeness of X, also convergent. Hence it follows for every $g \in D(A)$ that

$$\left\langle \lim_{\nu \to \infty} f_\nu - f | A g \right\rangle = \lim_{\nu \to \infty} \langle f_\nu | A g \rangle - \langle f | A g \rangle$$
$$= \lim_{\nu \to \infty} \langle A f_\nu | g \rangle - \langle f | \bar{A} g \rangle = \left\langle \bar{A}^* f | g \right\rangle - \langle f | \bar{A} g \rangle = 0$$

and hence, since Ran A is dense in X, that

$$\lim_{\nu \to \infty} f_\nu = f \; .$$

Hence it follows that $(f, \bar{A}^* f) \in G(\bar{A})$ and also (A.6). Finally, we conclude that $\bar{A}^* = \bar{A}$. '(ii)': For this, let $\gamma \in \mathbb{R}$ be such that

$$\langle f | A f \rangle \geqslant \gamma \, \langle f | f \rangle$$

for every $f \in D(A)$ and such that there is $\gamma' < \gamma$ such that $\mathrm{Ran}(A - \gamma')$ is dense in X. Then

$$\| f \| \cdot \| (A - \gamma') f \| \geqslant | \langle f | (A - \gamma') f \rangle | \geqslant (\gamma - \gamma') \| f \|^2$$

for every $f \in D(A)$ and hence

$$\| (A - \gamma') f \| \geqslant (\gamma - \gamma') \| f \|$$

for every $f \in D(A)$. Since $A - \gamma'$ is a densely-defined, linear and symmetric operator in X, it follows by (i) that $A - \gamma'$ is essentially self-adjoint. This implies that

$$\bar{A} - \gamma' = \overline{A - \gamma'} = (\overline{A - \gamma'})^* = (\bar{A} - \gamma')^* = \bar{A}^* - \gamma'$$

and hence that A is essentially self-adjoint. $\qquad\square$

Lemma A.3.6 *(**Rank-nullity theorem for linear operators**) Let $(X, \langle | \rangle_X)$ and $(Y, \langle | \rangle_Y)$ be Hilbert spaces over $\mathbb{K} \in \{\mathbb{R}, \mathbb{C}\}$, and A be a densely-defined and linear Y-valued operator in X. Then*

$$\ker A^* = (RanA)^\perp .$$

Proof '$\subset$': For this, let $g \in \ker A^*$. Then, it follows that

$$0 = \langle A^*g | f \rangle_X = \langle g | Af \rangle_Y$$

for all $f \in D(A)$ and hence that $g \in (RanA)^\perp$.
'$\supset$': For this, let $g \in (RanA)^\perp$. Then, it follows that

$$0 = \langle g | Af \rangle_Y$$

for all $f \in D(A)$ and hence that $g \in D(A^*)$ as well as that $A^*g = 0$.

$\qquad\square$

Theorem A.3.7 *(**A characterization of essential self-adjointness**) Let $(X, \langle | \rangle)$ be a complex Hilbert space and $A : D(A) \to X$ be a densely-defined, linear, symmetric operator in X. Then A is essentially self-adjoint if and only if $Ran(A - i)$ and $Ran(A + i)$ are dense in X.*

Proof For this, we note that for $\lambda \in \{-i, i\}$ and $f \in D(A)$ it follows that

$$\begin{aligned}
\|(A - \lambda)f\|^2 &= \langle Af - \lambda f | Af - \lambda f \rangle \\
&= \|Af\|^2 + |\lambda|^2 \|f\|^2 - \lambda^* \langle f | Af \rangle - \lambda \langle Af | f \rangle \\
&= \|Af\|^2 + |\lambda|^2 \|f\|^2 - \lambda^* \langle f | Af \rangle - \lambda \langle f | Af \rangle = \|f\|_A^2 \geqslant \|f\|^2
\end{aligned}$$

and hence that

$$\|(A - \lambda)f\| = \|f\|_A \geqslant \|f\| . \tag{A.7}$$

In particular, the last implies that

$$\ker(A - \lambda) = \{0\} .$$

'$\Rightarrow$': For this, let A be essentially self-adjoint and $\lambda \in \{-1, 1\}$. Then according to the Lemma A.3.6,

$$[\mathrm{Ran}(\bar{A} + \lambda)]^{\perp} = \ker(\bar{A} - \lambda) = \{0\} \ .$$

Since $\mathrm{Ran}(A + \lambda)$ is dense in $\mathrm{Ran}(\bar{A} + \lambda)$, the last also implies that

$$[\mathrm{Ran}(A + \lambda)]^{\perp} = \{0\}$$

and hence that

$$\overline{\mathrm{Ran}(A + \lambda)} = [\mathrm{Ran}(A + \lambda)]^{\perp\perp} = X \ .$$

'$\Leftarrow$': For this, let $\mathrm{Ran}(A - \lambda)$ be dense in X for $\lambda \in \{-i, i\}$. In a first step, we show, for $\lambda \in \{-i, i\}$, that

$$\mathrm{Ran}(\bar{A} - \lambda) = X \ . \tag{A.8}$$

Since $\mathrm{Ran}(A - \lambda)$ is dense in X, for $g \in X$, there is a sequence $f_1, f_2, \ldots$ of elements of $D(A)$ such that

$$\lim_{\nu \to \infty} (A - \lambda) f_\nu = g \ .$$

Hence it follows by (A.7) that $f_1, f_2, \ldots$ is a Cauchy sequence in X and hence, by the completeness of X, convergent to some $f \in X$. In particular, this implies that $f \in D(\bar{A})$ as well as that

$$(\bar{A} - \lambda) f = g \ .$$

Hence it follows (A.8). Since $\bar{A}$ is symmetric, i.e., $\bar{A}^* \supset \bar{A}$, for the proof of self-adjointness of $\bar{A}$, it is sufficient to show that $D(\bar{A}^*) \subset D(\bar{A})$. For this, let $f \in D(\bar{A}^*)$. Then according to (A.8), there is $g \in D(\bar{A})$ such that

$$(\bar{A} - \lambda) g = (\bar{A}^* - \lambda) f \ .$$

Hence it follows by Lemma A.3.6 that

$$f - g \in \ker(\bar{A}^* - \lambda) = [\mathrm{Ran}(\bar{A} + \lambda)]^{\perp} = \{0\}$$

and hence that $f = g \in D(\bar{A})$. $\square$

Corollary A.3.8 *(A characterization of self-adjointness, compare [6], Theorem 1.9.4)* *Let $(X, \langle | \rangle)$ be a complex Hilbert space and $A : D(A) \to X$ be a densely-defined, linear, symmetric operator in X. Then A is self-adjoint if and only if $\mathrm{Ran}(A - i) = X$ and $\mathrm{Ran}(A + i) = X$.*

Proof If A is self-adjoint, then A is essentially self-adjoint and closed. From Theorem A.3.7, we conclude that $\mathrm{Ran}(A - i)$ and $\mathrm{Ran}(A + i)$ are dense in X. According the proof of Theorem A.3.7, this implies that

$$\text{Ran}(A - \lambda) = \text{Ran}(\bar{A} - \lambda) = X$$

for $\lambda \in \{-i, i\}$. If $\text{Ran}(A - i) = X$ and $\text{Ran}(A + i) = X$, we conclude from Theorem A.3.7 and the corresponding proof that A is essentially self-adjoint as well as that $A - i$ and $A + i$ are bijective. As a consequence, $(A - i)^{-1}, (A + i)^{-1} \in L(X, X)$. Further, for $f \in D(\bar{A})$, there is a sequence $f_0, f_1, \ldots$ in $D(A)$ such that

$$\lim_{\nu \to \infty} f_\nu = f \ , \quad \lim_{\nu \to \infty} A f_\nu = \bar{A} f \ .$$

Hence also

$$\lim_{\nu \to \infty} (A - i) f_\nu = (\bar{A} - i) f$$

and

$$f = \lim_{\nu \to \infty} f_\nu = (A - i)^{-1} (\bar{A} - i) f \in D(A).$$

Therefore, it follows that $\bar{A} = A$ and that A is self-adjoint.   □

Corollary A.3.9 (*Compare [6], Theorem 1.9.3*) *Let* $(X, \langle \, | \, \rangle)$ *be a complex Hilbert space and* $A : D(A) \to X$ *be a densely-defined, linear, symmetric operator in* X.

(i) *If* $\text{Ran}(A^2 + 1)$ *is dense in* X, *then* A *is essentially self-adjoint.*
(ii) *If* A *is essentially self-adjoint, then* $\text{Ran}(\bar{A}^2 + 1)$ *is dense in* X.

Proof "(i)": First, we note that $f \in D(A^2 + 1)$ if and only if $f \in D(A)$ and $Af \in D(A)$. If $\text{Ran}(A^2 + 1)$ is dense in X, since

$$(A^2 + 1) f = (A - i)(A + i) f = (A + i)(A - i) f$$

for $f \in D(A^2 + 1)$, it follows that $\text{Ran}(A - i)$ and $\text{Ran}(A + i)$ are dense in X. Hence, according to Theorem A.3.7, A is essentially self-adjoint.
"(ii)": If A is essentially self-adjoint, $\bar{A}$ is self-adjoint and, for instance, according to the proof of Theorem A.3.7, $\bar{A} - i$ and $\bar{A} + i$ are bijective. Hence it follows for $g \in X$ that

$$g = (\bar{A} - i)(\bar{A} + i)(\bar{A} + i)^{-1}(\bar{A} - i)^{-1} g = (\bar{A}^2 + 1)(\bar{A} + i)^{-1}(\bar{A} - i)^{-1} g \ .$$

As consequence,

$$\text{Ran}(\bar{A}^2 + 1) = X \ .$$

□

A.3.4   A Symmetric Operator Without Self-Adjoint Extensions

An appropriate[5] choice for the state space/representation space, for a particle confined to the half-line $I := (0, \infty)$, is $X := L^2_\mathbb{C}(I)$. The representation turns into a position representation, by choosing as representation of the position operator the maximal multiplication operator in X that multiplies every element of its domain $D(\hat{q})$ by the function u/k on I, where u denotes the identical function on I and $\kappa > 0$ has dimension 1/length is a constant with dimension 1/length. As a maximal multiplication operator, $\hat{q}$ is densely-defined, linear and self-adjoint. The spectrum of $\hat{q}$ coincides with the essential range of u/κ, given by $(0, \infty)$. In the next step, before defining a quantization of the still unspecified Hamiltonian of the system, we need to represent the canonically conjugate momentum of the system in such a way that the momentum operator $\hat{p}$ and $\hat{q}$ satisfy the canonical commutation relations (CCR). Practically, this leaves as only option for $\hat{p}$ a self-adjoint extension of the densely-defined, linear and symmetric operator $\hat{p}_0$, with domain $D(\hat{p}_0)$ given by[6]

$$\{\, f \in C^1(I, \mathbb{C}) \cap L^2_\mathbb{C}(I) : f' \in L^2_\mathbb{C}(I) \wedge \lim_{u \to 0} f(u) = 0 \,\} \,,$$

where $C^1(I, \mathbb{C})$ denotes the vector space of continuously differentiable functions on I, and such that

$$\hat{p}_0 f := \frac{\hbar\kappa}{i}\, f' \,,$$

for every $f \in D(\hat{p}_0)$. It turns out that such an extension does not exist. In the following, we give a proof of this fact. For convenience, in this proof we multiply $\hat{p}_0$ by -1 and drop the factor $\hbar\kappa$. Both changes do not influence the result in a qualitative way.[7]

Theorem A.3.10 *(A symmetric operator without self-adjoint extensions) Let $A : D(A) \to L^2_\mathbb{C}(I)$ be the densely-defined, linear operator in $L^2_\mathbb{C}(I)$, where $I := (0, \infty)$, defined by*

$$Af := if'$$

for every $f \in D(A)$, where

$$D(A) := \{\, f \in C^1(I, \mathbb{C}) \cap L^2_\mathbb{C}(I) : f' \in L^2_\mathbb{C}(I) \wedge \lim_{u \to 0} f(u) = 0 \,\} \,.$$

Then A is symmetric, but has no self-adjoint extensions.

---

[5] As mentioned before, the representation space is largely arbitrary. Any 2 infinite dimensional separable Hilbert spaces are isomorphic by a unitary transformation. More relevant is the representation of the observables.

[6] Reducing the domain of $\hat{p}_0$ to $C^1_0(I, \mathbb{C})$ does not influence the subsequent result. It is not difficult to prove that the closure of the operator that is obtained in this way is an extension of $\hat{p}_0$.

[7] The main consequence of dimensional factors is to give spectral values the right physical dimensions.

Proof The proof proceeds in 3 steps.

(1) In a first step, we show that A is symmetric and that $\mathrm{Ran}(A - i)$ is not dense in X.
(2) In a second step, we show that $\bar{A} + i$ is bijective.
(3) In a third step, we conclude from (1) and (2) that there is no self-adjoint extension of A.

'(1)': First, we show an auxiliary result that for $f \in C^1(I, \mathbb{C}) \cap L^1_{\mathbb{C}}(I)$ such that[8] $f \in C(\bar{I}, \mathbb{C})$ and $f' \in L^1_{\mathbb{C}}(I)$, it follows that

$$\lim_{u \to \infty} f(u) = 0 .\tag{A.9}$$

For the proof, let such f be given. Then

$$f(u) = \lim_{u \to 0} f(u) + \int_0^u f'(v) \, dv . \tag{A.10}$$

For the proof of the latter, we define $h : I \to \mathbb{R}$ by

$$h(u) := \lim_{u \to 0} f(u) + \int_0^u f'(v) \, dv$$

for every $u \in I$. Then h is differentiable with derivative f' and hence, $f = h + C$ for some $C \in \mathbb{R}$. Since

$$\lim_{u \to 0} f(u) = \lim_{u \to 0} h(u),$$

it follows that $C = 0$ and hence the validity of (A.10). Hence, it follows the existence of

$$a := \lim_{u \to \infty} f(u) .$$

If $a \neq 0$, there is $u_0 > 0$ such that $|f(u)| > |a|/2$ for all $u \geqslant u_0$. Hence,

$$\int_0^u |f(v)| \, dv \geqslant \frac{|a|}{2}(u - u_0)$$

for every $u \in [u_0, \infty)$ which contradicts the existence of

$$\lim_{u \to \infty} \int_0^u |f(v)| \, dv .\notag$$

Hence, it follows the validity of (A.9). With the help of the previous auxiliary result, it follows for $f, g \in D(A)$ that

---

[8] Here $C(\bar{I}, \mathbb{C})$ denotes the vector space consisting of all continuous complex-valued on I that are restrictions of complex-valued continuous functions on $[0, \infty)$.

$$\langle f | Ag \rangle = i \int_0^\infty f^*(u)g'(u)\,du = i \int_0^\infty [\,(f^*g)'(u) - f'^*(u)g(u)\,]\,du = \langle Af | g \rangle$$

and hence that A is symmetric. On the other hand, again by use of the previous auxiliary result, it follows that

$$\langle e^{-u} | (A - i)f \rangle = i \int_0^\infty e^{-u}(f' - f)(u)\,du = i \int_0^\infty (e^{-u}f)'(u)\,du$$

$$= i \lim_{n \to \infty} [\,(e^{-u}f)(n) - (e^{-u}f)(1/n)\,] = 0 \;.$$

Hence $\mathrm{Ran}(A - i)$ is not dense in X, and therefore A is not essentially self-adjoint.
'(2)': For the proof that $\bar{A} + i$ is bijective, we define for every $g \in C_0(I, \mathbb{C})$ a corresponding $B_0 g$ by[9]

$$(B_0 g)(u) := -ie^{-u} \int_0^u e^v g(v)\,dv = -i \int_0^u e^{v-u} g(v)\,dv$$

for every $u \in I$. Then $B_0 g \in C^1(I, \mathbb{C})$ such that

$$i(B_0 g)' + i B_0 g = g \;, \quad \lim_{u \to 0} (B_0 g)(u) = 0$$

for every $u \in I$. Further, if $R > 0$ is such that $\mathrm{supp}(g) \subset [0, R]$, it follows that

$$|(B_0 g)(u)| \leqslant e^{-u} \left(\int_0^u \chi_{[0,R]}(v) \cdot e^{2v}\,dv \right)^{1/2} \|g\|_2 \leqslant \frac{1}{\sqrt{2}} \sqrt{e^{2R} - 1}\,\|g\|_2\, e^{-u} \;,$$

for every $u \in I$ and hence that $B_0 g \in X$ as well as that

$$\|B_0 g\|_2 \leqslant \frac{1}{\sqrt{2}} \sqrt{e^{2R} - 1}\,\|e^{-u}\|_2 \cdot \|g\|_2 \;.$$

As a consequence, by $B_0 := (C_0(I, \mathbb{C}) \to X, g \mapsto B_0 g)$, there is given a densely-defined, linear operator in X such that

$$(A + i)B_0 g = g \;, \quad \|g\|_2 = \|(A + i)B_0 g\|_2 \geqslant \|B_0 g\|_2$$

for every $g \in D(B_0)$. Hence B_0 is also bounded, and there is a unique extension of B_0 to a bounded linear operator B on X. Since $C_0(I, \mathbb{C})$ is dense in X, for every $g \in X$ there is a sequence $g_1, g_2, \ldots$ in $C_0(I, \mathbb{C})$ that is convergent to g. Also, since B is bounded, $B_0 g_1, B_0 g_2, \ldots$ is convergent to Bg and hence $A B_0 g_1 = g_1 - i B_0 g_1$, $A B_0 g_2 = g_2 - i B_0 g_2, \ldots$ convergent to $g - i Bg$. Therefore, $Bg \in D(\bar{A})$ and $(\bar{A} + i)Bg = g$. Hence $\bar{A} + i$ is surjective. Since, as consequence of the symmetry of $\bar{A}$, $\bar{A} + i$ is also injective, it follows that $\bar{A} + i$ is bijective.

---

[9] Here $C_0(I, \mathbb{C})$ denotes the vector space consisting of all continuous complex-valued functions f on I with a compact support that is contained in I, i.e., the closure of $\{u \in I : f(u) = 0\}$ is a compact subset of I, implying that $\lim_{u \to 0} f(u) = 0$.

'(3)': Finally, let $\hat{A}$ be a self-adjoint extension of A. Then $\hat{A} + i$ is a bijective linear and closed extension of $A + i$. Then $(\hat{A} + i)^{-1}$, $(\bar{A} + i)^{-1}$ are bounded linear operators that coincide on the dense domain, $\text{Ran}(A + i)$, and hence are identical. As a consequence, $\hat{A} = \bar{A}$. On the other hand, $\bar{A}$ is not self-adjoint. Hence there is no self-adjoint extension of A. □

In a further step, we are going to show that the spectrum $\sigma(\bar{A})$ of $\bar{A}$ is given by the closed lower half-plane of the complex plane, giving a further reason that $(-\hbar\kappa)\bar{A}$ cannot be used as a representation of a physical observable.

Corollary A.3.11 *Let A be defined as in Theorem A.3.10. The spectrum $\sigma(\bar{A})$ of $\bar{A}$ is given by the closed upper half-plane of the complex plane,*

$$\sigma(\bar{A}) = \mathbb{R} \times [0, \infty) \ .$$

Proof For the proof, let $\lambda \in (0, \infty) \times \mathbb{R}$. We note that this implies that $i\lambda^* \in \mathbb{R} \times (0, \infty)$. It follows for $f \in D(A)$ that

$$\left\langle e^{-\lambda u} | (A - i\lambda^*) f \right\rangle = i \int_0^\infty e^{-\lambda^* u} (f' - \lambda^* f)(u) \, du = i \int_0^\infty (e^{-\lambda^* u} f)'(u) \, du$$

$$= i \lim_{n \to \infty} [\, (e^{-\lambda^* u} f)(n) - (e^{-\lambda^* u} f)(1/n) \,] = 0 \ .$$

Further, the same result is true for every $f \in D(\bar{A})$, since for each $f \in D(\bar{A})$, there is a sequence $f_1, f_2, \ldots$ in $D(A)$ such that

$$\lim_{\nu \to \infty} f_\nu = f$$

and at the same time such that

$$\lim_{\nu \to \infty} A f_\nu = \bar{A} f \ .$$

Hence it follows that

$$\text{Ran}(\bar{A} - i\lambda^*)$$

is not dense in X. As a consequence, $\bar{A} - i\lambda^*$ is not bijective and $i\lambda^* \in \sigma(\bar{A})$. Thus, $\mathbb{R} \times (0, \infty) \subset \sigma(\bar{A})$ and, since $\sigma(\bar{A})$ is closed, that

$$\mathbb{R} \times [0, \infty) \subset \sigma(\bar{A}) \ .$$

For the next step, let $\lambda \in (-\infty, 0) \times \mathbb{R}$. This implies that $i\lambda^* \in \mathbb{R} \times (-\infty, 0)$. For the proof that $\bar{A} - i\lambda^*$ is bijective, we define for every $g \in C_0(I, \mathbb{C})$ a corresponding $B_0 g$ by

$$(B_0 g)(u) := -i e^{\lambda^* u} \int_0^u e^{-\lambda^* v} g(v) \, dv = \int_0^u [-i e^{-\lambda^*(v-u)}] g(v) \, dv$$

for every $u \in I$, where

$$K(u, v) := \begin{cases} -ie^{-\lambda^*(v-u)} & 0 < v < u \\ 0 & u \leqslant v \end{cases} .$$

Then $B_0 g \in C^1(I, \mathbb{C})$ such that

$$i(B_0 g)' - i\lambda^* B_0 g = g , \quad \lim_{u \to 0} (B_0 g)(u) = 0$$

for every $u \in I$. In addition, for every $u \in I$, $K(u, \cdot)$ is integrable such that

$$\int_0^\infty |K(u, v)| \, dv = \int_0^u e^{|\mathrm{Re}(\lambda)|(v-u)} \, dv$$
$$= \frac{1}{|\mathrm{Re}(\lambda)|} e^{-|\mathrm{Re}(\lambda)|u} \left(e^{|\mathrm{Re}(\lambda)|u} - 1 \right) \leqslant \frac{1}{|\mathrm{Re}(\lambda)|} ,$$

and for every $v \in I$, $K(\cdot, v)$ is integrable such that

$$\int_0^\infty |K(u, v)| \, du = \int_v^\infty e^{|\mathrm{Re}(\lambda)|(v-u)} \, du = \frac{1}{|\mathrm{Re}(\lambda)|} .$$

Hence B_0 is bounded with operator norm

$$\|B_0\| \leqslant \frac{1}{|\mathrm{Re}(\lambda)|} .$$

As a consequence, by $B_0 := (C_0(I, \mathbb{C}) \to X, g \mapsto B_0 g)$, there is given a densely-defined, linear operator in X such that

$$(A - i\lambda^*) B_0 g = g .$$

Since B_0 is also bounded, there is a unique extension of B_0 to a bounded linear operator B on X. This operator B is given by the integral operator $\mathrm{Int} K$ on X, with kernel function K. In particular,

$$\|B\| \leqslant \frac{1}{|\mathrm{Re}(\lambda)|} .$$

Since $C_0(I, \mathbb{C})$ is dense in X, for every $g \in X$ there is a sequence $g_1, g_2, \ldots$ in $C_0(I, \mathbb{C})$ that is convergent to g. Also, since B is bounded, $B_0 g_1, B_0 g_2, \ldots$ is convergent to Bg and hence $AB_0 g_1 = g_1 + i\lambda^* B_0 g_1$, $AB_0 g_2 = g_2 + i\lambda^* B_0 g_2, \ldots$ convergent to $g + i\lambda^* Bg$. Therefore, $Bg \in D(\bar{A})$ and $(\bar{A} - i\lambda^*) Bg = g$. Hence $\bar{A} - i\lambda^*$ is surjective. Also, we note for $f \in D(A)$ that

$$[B(A - i\lambda^*)f](u) = -ie^{\lambda^* u} \int_0^u e^{-\lambda^* v}[(A - i\lambda^*)f](v)\, dv$$

$$= e^{\lambda^* u} \int_0^u e^{-\lambda^* v}[f'(v) - \lambda^* f(v)]\, dv = e^{\lambda^* u} \int_0^u (e^{-\lambda^* v} f)'(v)\, dv$$

$$= e^{\lambda^* u} e^{-\lambda^* u} f(u) = f(u)\ ,$$

for every $u \in I$. Hence,

$$B(A - i\lambda^*)f = f\ ,$$

for every $f \in D(A)$. Further, for every $f \in D(\bar{A})$, there is a sequence $f_1, f_2, \ldots$ in $D(A)$ such that

$$\lim_{\nu \to \infty} f_\nu = f\ ,\quad \lim_{\nu \to \infty} f_\nu = \bar{A} f\ .$$

Hence,

$$\lim_{\nu \to \infty} (A - i\lambda^*)f_\nu = (\bar{A} - i\lambda^*)f\ ,$$

and

$$f = \lim_{\nu \to \infty} f_\nu = \lim_{\nu \to \infty} B(A - i\lambda^*)f_\nu = B(\bar{A} - i\lambda^*)f\ .$$

As a consequence, it follows that

$$B(\bar{A} - i\lambda^*)f = f\ ,$$

for every $f \in D(\bar{A})$. The latter implies that $\bar{A} - i\lambda^*$ is also injective. Finally, it follows that $\bar{A} - i\lambda^*$ is bijective, which implies that $i\lambda^* \notin \sigma(\bar{A})$ and hence that

$$[\mathbb{R} \times (-\infty, 0)] \cap \sigma(\bar{A}) = \emptyset\ .$$

$$\square$$

References

1. Reed M and Simon B, 1980, 1975, 1979, 1978, *Methods of modern mathematical physics*, Volume I, II, III, IV, Academic: New York.
2. Riesz F and Sz-Nagy B 1955, *Functional analysis*, Unger: New York.
3. Rudin W 1991, *Functional analysis*, 2nd ed., MacGraw-Hill: New York.
4. Weidmann J 1980, *Linear Operators in Hilbert spaces*, Springer: New York.
5. Yosida K 1968, *Functional analysis*, 2nd ed., Springer: Berlin.
6. Schechter M 2003, *Operator methods in quantum mechanics*, Dover Publication: New York.
7. Adams R A, Fournier J J F 2003, *Sobolev spaces*, 2nd ed., Academic Press: New York.
8. Brezis H 1983, *Analyse fonctionnelle: Théorie et applications*, Collection Mathématiques Appliquées pour la Maîtrise, Masson: Paris.
9. Cohen-Tannoudji C, Diu B, Laloe F 1978, *Quantum Mechanics*, Volume 1, Wiley & Sons: New York.

10. Dirac P A M 2019, *The principles of quantum mechanics*, BN Publishing: New York.
11. Dunford N, Schwartz J T 1963, *Linear operators, Part II: Spectral theory: Self adjoint operators in Hilbert space theory*, Wiley: New York.
12. Heisenberg W 1925, *Über quantentheoretische Umdeutung kinematischer und mechanischer Beziehungen.* Zeitschrift für Physik, Vol. **33**, 879–893.
13. Hirzebruch F, Scharlau W 1971, *Einführung in die Funktionalanalysis*, BI: Mannheim.
14. Landau L D, Lifshitz E M 1991, 3rd Ed, *Quantum mechanics: Non-relativistic theory*, Pergamon Press: Oxford.
15. Mackey G W 2004, *Mathematical foundations of quantum mechanics*, Dover Publications: Dover.
16. Mermin D 2004, *What is wrong with this pillow*, Nature, Vol. **505**, 153–155.
17. Messiah A 2014, *Quantum mechanics*, Dover Publication: New York.
18. Meyenn K 1996, *Wolfgang Pauli: Scientific Correspondence with Bohr, Einstein, Heisenberg, a.o.*, Volume IV, Part I: 1950 – 1952, Springer: Berlin.
19. Von Neumann J 1932, *Mathematische Grundlagen der Quantenmechanik*, Springer: Berlin.
20. Prugovecki E 1981, *Quantum Mechanics in Hilbert Space*, Academic Press: New York.
21. Renardy M and Rogers R C 1993, *An introduction to partial differential equations*, Springer: New York.
22. Sakurai J J 1967, *Advanced quantum mechanics*, Addison-Wesley: Reading.
23. Schiff L I 1968, *Quantum mechanics*, 3Rev Ed, McGraw-Hill Education: New York.
24. Stein E M, Shakarchi R 2003, *Fourier analysis: An introduction*, Princeton University Press: Princeton and Oxford.
25. Thirring W 1981, *A Course in Mathematical Physics 3: Quantum Mechanics of Atoms and Molecules*, Springer: New York.
26. Ziemer W P 1989, *Weakly differentiable functions*, Springer: New York.

Index

H. Beyer, *The Reasoning of Quantum Mechanics*, Synthesis Lectures on Engineering,
Science, and Technology, https://doi.org/10.1007/978-3-031-17177-2

Symbol Index

© The Editor(s) (if applicable) and The Author(s), under exclusive license
to Springer Nature Switzerland AG 2022
H. Beyer, *The Reasoning of Quantum Mechanics*, Synthesis Lectures on Engineering,
Science, and Technology, https://doi.org/10.1007/978-3-031-17177-2

Printed in the United States
by Baker & Taylor Publisher Services